Allocating Federal Funds for State Programs for English Language Learners

Panel to Review Alternative Data Sources for the Limited-English Proficiency Allocation Formula under Title III, Part A, Elementary and Secondary Education Act

Committee on National Statistics
and
Board on Testing and Assessment

Division of Behavioral and Social Sciences and Education

NATIONAL RESEARCH COUNCIL
OF THE NATIONAL ACADEMIES

THE NATIONAL ACADEMIES PRESS
Washington, D.C.
www.nap.edu

THE NATIONAL ACADEMIES PRESS 500 Fifth Street, N.W. Washington, DC 20001

NOTICE: The project that is the subject of this report was approved by the Governing Board of the National Research Council, whose members are drawn from the councils of the National Academy of Sciences, the National Academy of Engineering, and the Institute of Medicine. The members of the committee responsible for the report were chosen for their special competences and with regard for appropriate balance.

This study was supported by U.S. Department of Education Contract Number ED-08-CO-0119 to the National Academy of Sciences. Support of the work of the Committee on National Statistics is provided by a consortium of federal agencies through a grant from the National Science Foundation (award number SES-0453930). Any opinions, findings, conclusions, or recommendations expressed in this publication are those of the author(s) and do not necessarily reflect the views of the organizations or agencies that provided support for the project.

International Standard Book Number-13: 978-0-309-18658-2
International Standard Book Number-10: 0-309-18658-7

Additional copies of this report are available from the National Academies Press, 500 Fifth Street, N.W., Lockbox 285, Washington, DC 20055; (800) 624-6242 or (202) 334-3313 (in the Washington metropolitan area); Internet, http://www.nap.edu

Printed in the United States of America

Suggested citation: National Research Council. (2011). *Allocating Federal Funds for State Programs for English Language Learners.* Panel to Review Alternative Data Sources for the Limited-English Proficiency Allocation Formula under Title III, Part A, Elementary and Secondary Education Act. Committee on National Statistics and Board on Testing and Assessment. Division of Behavioral and Social Sciences and Education. Washington, DC: The National Academies Press.

THE NATIONAL ACADEMIES

Advisers to the Nation on Science, Engineering, and Medicine

The **National Academy of Sciences** is a private, nonprofit, self-perpetuating society of distinguished scholars engaged in scientific and engineering research, dedicated to the furtherance of science and technology and to their use for the general welfare. Upon the authority of the charter granted to it by the Congress in 1863, the Academy has a mandate that requires it to advise the federal government on scientific and technical matters. Dr. Ralph J. Cicerone is president of the National Academy of Sciences.

The **National Academy of Engineering** was established in 1964, under the charter of the National Academy of Sciences, as a parallel organization of outstanding engineers. It is autonomous in its administration and in the selection of its members, sharing with the National Academy of Sciences the responsibility for advising the federal government. The National Academy of Engineering also sponsors engineering programs aimed at meeting national needs, encourages education and research, and recognizes the superior achievements of engineers. Dr. Charles M. Vest is president of the National Academy of Engineering.

The **Institute of Medicine** was established in 1970 by the National Academy of Sciences to secure the services of eminent members of appropriate professions in the examination of policy matters pertaining to the health of the public. The Institute acts under the responsibility given to the National Academy of Sciences by its congressional charter to be an adviser to the federal government and, upon its own initiative, to identify issues of medical care, research, and education. Dr. Harvey V. Fineberg is president of the Institute of Medicine.

The **National Research Council** was organized by the National Academy of Sciences in 1916 to associate the broad community of science and technology with the Academy's purposes of furthering knowledge and advising the federal government. Functioning in accordance with general policies determined by the Academy, the Council has become the principal operating agency of both the National Academy of Sciences and the National Academy of Engineering in providing services to the government, the public, and the scientific and engineering communities. The Council is administered jointly by both Academies and the Institute of Medicine. Dr. Ralph J. Cicerone and Dr. Charles M. Vest are chair and vice chair, respectively, of the National Research Council.

www.national-academies.org

PANEL TO REVIEW ALTERNATIVE DATA SOURCES FOR THE LIMITED-ENGLISH PROFICIENCY ALLOCATION FORMULA UNDER TITLE III, PART A, ELEMENTARY AND SECONDARY EDUCATION ACT

Alan Zaslavsky (*Chair*), Department of Health Care Policy, Harvard Medical School
Jamal Abedi, School of Education, University of California, Davis
Frank D. Bean, Center for Research on Immigration, Population and Public Policy, University of California, Irvine
David Francis, Texas Institute for Measurement, Evaluation and Statistics, University of Houston
Edward Haertel, School of Education, Stanford University
David Hubble, WESTAT, Inc., Rockville, MD
Rebecca Kopriva, Wisconsin Center for Education Research and School of Education, University of Wisconsin, Madison
Robert Linquanti, WestEd, Oakland, CA
Helen Malagon, Office of Superintendent of Public Education, State of Washington Migrant and Bilingual Education Programs
Catherine Neff, Office of Federal Programs, South Carolina Department of Education
P. David Pearson, Graduate School of Education, University of California, Berkeley
Charlene Rivera, Graduate School of Education and Center for Equity and Excellence in Education, George Washington University

Judith A. Koenig, *Costudy Director*
Thomas Plewes, *Costudy Director*
Esha Sinha, *Research Associate*
Michael J. Siri, *Program Associate*

COMMITTEE ON NATIONAL STATISTICS
2010

Acknowledgments

This report is the product of the work by many people, each of whom contributed expertise to the enterprise. I first acknowledge with great appreciation the efforts of my fellow panel members, who brought expertise as data users and state data providers and in education policy, demography, statistical estimation methods, census and American Community Survey (ACS) methodology, administrative data systems, and testing and assessment. All of this expertise was critical to the multifaceted issues that had to be addressed in this evaluation of the admissible data sources for allocation of funds under Title III, Part A, of the Elementary and Secondary Education Act. The panel members freely contributed their time to accomplishing the myriad of tasks associated with assembling information and preparing this report and cheerfully accepted the duties of facilitating sessions at the five committee meetings. Their contributions during the period in which the report was in final preparation and after the outside review, when sections of the report had to be turned around on a very truncated schedule, are especially appreciated. These efforts manifested the panel members' deep dedication to the well-being and effective education of English language learners and immigrant children.

The panel held public sessions at its first and second meetings. The sessions were organized as formal workshops, with presentations by internal and outside presenters followed by free-flowing discussion. In developing these public sessions, the committee greatly benefitted from the support of the staff of the U.S. Department of Education and the U.S. Census Bureau.

The panel was set on the right path at its first open meeting by Thomas M. Corwin, director, Division of Elementary, Secondary and Vocational Analysis, Budget Service, U.S. Department of Education, and Richard Smith, the department's acting assistant deputy secretary and director, Office of English Language Acquisition. They elaborated on the department's charge to the committee and established the background for the study for the information of the panel.

This report was preceded by a significant report by the U.S. Government Accountability Office in 2006 that described the allowable data for the allocation of formula-based grants to states to support the education of students with limited English proficiency. It described the effects of using the ACS and the states' assessment data, and it recommended a series of steps for the U.S. Department of Education to improve the quality and use of the data. The report's authors, Harriet C. Ganson, assistant director for education, workforce and income security, and Nagla'a El-Hodiri, senior economist, discussed the findings of that report with the panel at its April 2009 workshop.

The panel was assisted in developing a full understanding of the ACS at its first meeting by Susan Schecter, chief of the American Community Survey Office of the U.S. Census Bureau. She updated the committee on the current status of the survey and discussed the plans for development of estimates for small areas and populations.

At its second meeting, the panel heard from experts on state testing, state practices, and programs for English language learners, as well as the U.S. Department of Education official responsible for the development and maintenance of the departmental databases that hold the state data. We thank a group of Title III directors for their contribution at the workshop: Celina Arias-Romero (California), committee member Helen Malagon (Washington), Joanne Marino (North Carolina), Barbara Medina (Colorado), Elizabeth Minjarez (Texas), committee member Catherine Neff (South Carolina), Redro Ruiz (New York), and Maria Santos (New York City). The administrative data are strongly affected by state practices, and the members of this group documented the state practices that reflect the individual policies and needs as perceived by the state agencies that conduct these programs.

Ross Santy, director of the Performance Information Management Service in the Office of Planning, Evaluation and Policy Development, U.S. Department of Education, provided a summary of the types of data that are maintained on state programs, and traced the recent efforts to build a data system that increasingly standardizes and validates data maintained and submitted by state education agencies. His unit provided the committee with a number of special tabulations of data submitted by state education agencies to the U.S. Department of Education that greatly assisted our work.

Robert Kominski, assistant chief, Housing and Household Economic Statistics Division, U.S. Census Bureau, discussed the evolution and status of the questions on the ACS that provide the basic information for derivation of estimates of the number of limited English proficient students by state. The committee also extends special thanks to David Raglin and Sandra Clark of the Special Studies staff of the American Community Survey Office of the U.S. Census Bureau for facilitating the provision of extensive special tabulations of ACS data that permitted the panel to assess the adequacy of the ACS for allocation purposes.

We also acknowledge the guidance and support provided to this study by the two contracting officer technical representatives who were assigned to shepherd this project by the U.S. Department of Education. On numerous occasions, Pat Butler and Sandra Furey stepped in to assist in setting up meetings with the appropriate departmental experts, securing data for comparison purposes, and facilitating necessary contractual modifications to enable the successful completion of this study. Britt Jung of the department's budget office provided advice and guidance throughout the study and assisted in the provision of special funding for the purchase of ACS tabulations from the Census Bureau that permitted much of the analysis contained in this report. William Sonnenberg of the National Center for Education Statistics provided significant assistance to the panel about data used in the current and past allocation formula for the distribution of Title III funds by the U.S. Department of Education.

The committee gratefully acknowledges the dedicated effort provided by the staff of the Committee on National Statistics (CNSTAT) and the Board on Testing and Assessment (BOTA) of the Division of Behavioral and Social Sciences and Education of the National Research Council (NRC). Tom Plewes and Judy Koenig brought deep experience and broad organizational skill to their service as costudy directors, and their work was a key factor in ensuring the efficient completion of this report. Esha Sinha of the CNSTAT staff supported the analysis by developing the research databases of the data provided in various formats by the U.S. Department of Education and the U.S. Census Bureau, and she managed, through extraordinary efforts, to conduct analyses for the committee. Michael Siri, also of the CNSTAT staff, provided smooth administrative management of the five meetings held in several venues. The task of contacting state directors of English language learner programs to ascertain the latest information on their screening and assessment tests was ably and cheerfully handled by Rose Neugroschel of BOTA. We also want to acknowledge the active participation and guidance provided by Constance Citro, director of CNSTAT, and Stuart Elliott, director of BOTA, whose advice and leadership were instrumental in moving this study from the planning to report preparation stages.

This report has been reviewed in draft form by individuals chosen for their diverse perspectives and technical expertise, in accordance with procedures approved by the Report Review Committee of the NRC. The purpose of this independent review is to provide candid and critical comments that assist the institution in making its report as sound as possible, and to ensure that the report meets institutional standards for objectivity, evidence, and responsiveness to the study charge. The review comments and draft manuscript remain confidential to protect the integrity of the deliberative process.

The panel thanks the following individuals for their review of this report: Jonathan G. Dings, Planning and Assessment, Boulder Valley School District; Gerunda B. Hughes, Office of Institutional Assessment and Evaluation, Howard University; Joseph B. (Jay) Kadane, Department of Statistics, Carnegie Mellon University; Neal Kingston, Center for Educational Testing and Evaluation, University of Kansas; Jan Lanier, Division of College and Career Readiness, Tennessee Department of Education; Lilia G. Sánchez, English Learner and Curriculum Support Division,

California Department of Education; and Robin Scarcella, Academic English/ESL Program, University of California, Irvine. Although the reviewers listed above have provided many constructive comments and suggestions, they were not asked to endorse the conclusions or recommendations, nor did they see the final draft of the report before its release. The review of this report was overseen by Allen L. Schirm, Human Services Research, Mathematica Policy Research, Inc., and Kenneth W. Wachter, Demography and Statistics, University of California, Berkeley. Appointed by the NRC, they were responsible for making certain that the independent examination of this report was carried out in accordance with institutional procedures and that all review comments were carefully considered. Responsibility for the final content of the report rests entirely with the authoring committee and the NRC.

Alan Zaslavsky, *Chair*
Panel to Review Alternative Data Sources
for the Limited-English Proficiency
Allocation Formula under Title III, Part A,
Elementary and Secondary Education Act

Contents

Tables, Figures, and Boxes

TABLES

FIGURES

BOXES

Acronyms and Abbreviations

ACCESS	Assessing Comprehension and Communication in English State to State
ACS	American Community Survey
AERA	American Educational Research Association
AIR	American Institutes for Research
AMAO	annual measurable achievement objective
APA	American Psychological Association
AYP	adequate yearly progress
AZELLA	Arizona English Language Learner Assessment
BOTA	Board on Testing and Assessment
C2SS	Census 2000 Supplementary Survey
CAPI	computer-assisted personal interviewing
CATI	computer-assisted telephone interviewing
CCD	Common Core of Data
CCSSO	Council of Chief State School Officers
CDE	California Department of Education
CELA	Colorado English Language Assessment
CELDT	California English Language Development Test
CELLA	Comprehensive English Language Learning Assessment
CMA for ELA	California Modified Assessment for English-Language Arts
CNSTAT	Committee on National Statistics
CRESST	National Center for Research on Evaluation, Standards and Student Testing

CSAP	Colorado Student Assessment Program
C-SAVE	Center for the Study of Assessment Validity and Evaluation
CSPR	Consolidated State Performance Report
CST for ELA	California Standards Test for English-Language Arts
CV	coefficients of variation
DoEd	U.S. Department of Education
EDEN	Education Data Exchange Network
ELDA	English Language Development Assessment
ELL	English language learner
ELP	English language proficiency
ELPAS	English Language Proficiency Assessment Standards
ELPS	English Language Proficiency Survey
EPAS	English Proficiency for All Students
ESEA	Elementary and Secondary School Act
ESL	English as a second language
ESOL	English for speakers of other languages
ESS	EDEN Submission System
ETS	Educational Testing Service
GAO	U.S. Government Accountability Office
HLS	home language survey
IELA	Idaho English Language Assessment
IEP	Immigrant Education Program
IPT	IDEA Proficiency Test
KELPA	Kansas English Language Proficiency Assessment
LAB-R	Language Assessment Battery-Revised
LAS	Language Assessment Scales
LAS Links	Language Assessment Scales Links
LEA	local education agency
LEP	limited English proficiency
LEP-SCASS	Limited English Proficient State Collaborative on Assessment and Student Standards
LIEP	Language Instruction Educational Program
LPTS	Language Proficiency Test Series
MAC II	Maculaitis Assessment of Competencies Test of Language Proficiency
MAD	mean absolute difference

MAF	master address file
MARD	mean absolute relative difference
MELA-O	Massachusetts English Language Assessment-Oral
MEPA-R/W	Massachusetts English Proficiency Assessment-Reading and Writing
MI-ELPA	Michigan English Language Proficiency Assessment
MN-SOLOM	Minnesota Modified Student Oral Language Observation Matrix
MontCAS	Montana Comprehensive Assessment System
MWAC	Mountain West Assessment Consortium
NCELA	National Clearinghouse for English Language Acquisition
NCES	National Center for Education Statistics
NCLB	No Child Left Behind Act
NCME	National Council on Measurement in Education
NGA Center	National Governors Association's Center for Best Practices
NRC	National Research Council
NV-ELPA	Nevada State English Language Proficiency Assessment
NYSED	New York State Education Department
NYSESLAT	New York State English as a Second Language Achievement Test
OCR	Office of Civil Rights, U.S. Department of Education
OELA	Office of English Language Acquisition, U.S. Department of Education
OR-ELPA	Oregon English Language Proficiency Assessment
OTELA	Ohio Test of English Language Acquisition
PEP	Population Estimates Program
PUMA	public-use microdata area
PUMS	public-use microdata samples
RFEP	reclassified as fluent English proficient
RMSE	root mean square residual error
RPTE	Reading Proficiency Tests in English
SAIPE	Small Area Income and Poverty Estimates Program
SEA	state education agency
SELP	Stanford English Language Proficiency Test
TEA	Texas Education Agency
TEAE	Test of Emerging Academic English
TELPA	Tennessee English Language Placement Assessment
TELPAS	Texas English Language Proficiency Assessment Systems
TESOL	Teachers of English to Speakers of Other Languages

UALPA	Utah Academic Language Proficiency Assessment
WESTELL	West Virginia Test for English Language Learners
WIDA	World-Class Instructional Design and Assessment
WLPT-II	Washington Language Proficiency Test II

Summary

As the United States continues to be a nation of immigrants and their children, the nation's school systems face increased enrollments of students whose primary language is not English. With the 2001 reauthorization of the Elementary and Secondary Education Act (ESEA) through the No Child Left Behind Act (NCLB), the allocation of federal funds for programs to assist these students to be proficient in English became formula-based: 80 percent on the basis of the population of children with limited English proficiency[1] and 20 percent on the basis of the population of recently immigrated children and youth.

Title III of ESEA directs the U.S. Department of Education (DoEd) to allocate funds based on the data that will yield the most accurate, up-to-date numbers, and further directs the secretary to choose the more accurate source of data between the two allowable data sources: the number of students assessed for English proficiency as reported to the federal government by each state education agency or data from the American Community Survey (ACS). The department determined that the ACS estimates are more accurate, and since 2005, those data have been the basis for the federal distribution of Title III funds.

Subsequently, an analysis of the two data sources by the U.S. Government Accountability Office (2006b) raised concerns about the fact that the data sources would allocate quite different amounts to the states and suggested that the DoEd develop and implement a transparent methodology for determining the relative accuracy of the two allowable data sources. In addition, while shortcomings were

[1]Legislative language generally uses the term "limited English proficiency" (LEP) to describe these children, while the education field generally uses the term "English language learner" (ELL). ELL is intended to emphasize that these children are learning English as a new language as they also acquire proficiency in academic subject matter and is the term used in this report.

noted in the data provided by the states, the ACS estimates were shown to fluctuate between years, causing concern among the states about the unpredictability and unevenness of program funding.

In this context, the DoEd commissioned the National Research Council's Committee on National Statistics and Board on Testing and Assessment to convene this panel to address the accuracy of the estimates from the two data sources and the factors that influence the estimates and to consider means of increasing the accuracy of the data sources or alternative data sources that could be used for allocation purposes.

DESIRABLE CHARACTERISTICS FOR TITLE III ALLOCATION FORMULA DATA

The panel developed a set of criteria for evaluating which of the allowable data sources the department should use for allocating federal Title III funds: conceptual fit, level of geographic fit, timeliness, quality, cost, fairness, stability, insensitivity to policy and methodological differences, transparency, and comparability.

- **Conceptual Fit.** A data element used in an allocation formula should meet the conceptual objectives of the program for which the allocation is aimed. In the case of allocating Title III education funds to states, a data element with a good conceptual fit is one that meets the definition provided in the legislation—the number of LEP and immigrant children and youth in a state. In a larger sense, however, considering the overall objective of the allocation of federal funds, a conceptually fitting data element would provide subnational governments a level of federal funding that is proportional to their need and circumstances. The ACS defines need as the numbers of children and youth who are eligible on the basis of their skill in speaking the English language. The state-provided counts are more conceptually fit in that they define need as those identified by schools through questionnaires and assessments, which are increasingly standardized.
- **Level of Geographic Detail.** The state is the key level of detail for which the data should be available. Both the ACS and state-provided estimates are available for both states and local education agencies (LEAs).
- **Timeliness.** The elapsed time between the reference period for the estimates and the period for which the allocations are being made should be as short as possible. Both the ACS and the state-provided estimates are available approximately 9 months following the reference period.
- **Quality.** The ACS data meet statistical reliability standards and are of acceptable precision. The state-provided counts of English language learners, based on administrative data, are not subject to sampling error, but there may be some different interpretation of the instructions for data collection. The state-provided counts of immigrant children and youth rely on LEA judgments and fall short of the quality of ACS estimates.

- **Cost.** Both ACS estimates and state-provided counts are available at minimal extra cost.
- **Fairness.** The U.S. Census Bureau, which administers the ACS, ensures the data are free from manipulation. State-provided counts have improved such that data are similarly free from manipulation; however, states' discretion over various policies may affect perceptions of fairness.
- **Stability.** The state-provided counts are relatively stable from year to year. The 1-year ACS estimates for smaller states are subject to noticeable variation. The 3-year ACS estimates are more stable than the 1-year estimates and than the state counts.
- **Insensitivity to Policies and Methodological Differences.** The ACS estimates are not sensitive to administrative practices or policy differences, but they may be sensitive to differences in the demographic composition of respondents. The state-provided counts are somewhat sensitive to state decisions regarding identification, testing, and program entry and exit policies.
- **Transparency.** The ACS data are collected by professional staff using highly standardized, well-documented methods. The state-provided counts are collected by varying methods and depend on LEA implementation so that documentation is not readily available.
- **Comparability.** The ACS data are comparable across geographic and demographic dimensions. The state-provided counts conform to basic federal definitions but are not comparable in their constructs due to differences in state tests and the states' classification and reclassification criteria.

In sum, the panel concludes that the ACS data are slightly better on some characteristics, while the state-provided data have advantages on others, which is not reflected in the current formula.

RECOMMENDATIONS

Data Sources for Title III Allocation Formula

RECOMMENDATION 1 As soon as technically possible, the U.S. Department of Education should begin to incorporate state-provided counts of English language learner students into Title III formula allocation calculations. Initially, the state-provided data should be given a weight of 25 percent of the allocation, with the remaining 75 percent weight given to the American Community Survey data.

RECOMMENDATION 2 In the portion of the allocation that is based on state-provided data, the U.S. Department of Education should use the state-provided count of the number of students who are determined not

to be English proficient on the basis of the state's English language proficiency test.

RECOMMENDATION 3 The U.S. Department of Education should continue to use the American Community Survey estimate as the basis for allocating the 20 percent of the Title III funds that are to be based on the population of recently immigrated children and youth (relative to national counts of these populations).

Research

RECOMMENDATION 4 The U.S. Census Bureau should conduct research on the accuracy of the American Community Survey language item for assessing population prevalence of English language learner children and youth, including the strength of its association with more comprehensive English language proficiency measures. With the objective of evaluating and improving the item, researchers should examine the effects on responses of situational, cultural, demographic, and socioeconomic factors, placement of the item in the questionnaire, and the ability of adult responders to make English language proficiency distinctions.

Weighting of Data Sources

RECOMMENDATION 5 When the quality and cross-state comparability of state-provided data have reached an acceptable standard, the weight given to the state-provided counts should be adjusted upward to the point at which the American Community Survey estimates and the state-provided counts contribute equally to the 80 percent portion of the allocation formula. State-provided counts should continue to be based on the number of students who are determined not to be English proficient on the basis of the state's English language proficiency test, in a way that is comparable across states.

1

Introduction

The No Child Left Behind Act (NCLB) of 2001, which amended Title III of the Elementary and Secondary Education Act (ESEA), fundamentally changed how the federal government directs federal funding to support programs for children of limited English proficiency (LEP), also known as English language learner (ELL) students.[1] In the words of the U.S. Department of Education (DoEd) (2005b, p. 1), the NCLB "reflects a fundamental transformation in the relationship between the federal government and the states with regard to the education of LEP students."

THE POPULATION AND THE NEW LANDSCAPE

At its broadest level, an ELL student is one who has limited proficiency in the English language. Indeed, the ESEA provides a very specific definition of "limited English proficiency": see Box 1-1.

The legal definition includes elements that are relatively objective and those that are relatively subjective. The objective criteria cover demographics, background, and ability to meet the state's proficient level of achievement on state assessments; the subjective criteria cover perceived difficulties in sufficient command of the English language to be successful in classrooms in which the instructional language

[1]In this report, the committee uses both ELL and LEP students to describe the population of interest. The committee favors the term ELL as more descriptive of the population and the challenges that the population faces, but it recognizes that LEP is defined and used in the ESEA legislation and for Title III reporting purposes. By official definition, LEP students are "ages 3-21, enrolled in elementary or secondary education, born outside of the United States or speaking a language other than English in their homes, and not having sufficient mastery of English to meet state standards and excel in an English-language classroom" (Title III of ESEA).

BOX 1-1
Limited English Proficiency (LEP) Student: Definition

An LEP student is classified as one: (A.) who is aged 3 through 21; (B.) who is enrolled or preparing to enroll in an elementary school or secondary school; (C.) (i.) who was not born in the United States or whose native language is a language other than English; and who comes from an environment where a language other than English is dominant OR (ii.) (I.) who is a Native American or Alaska Native, or a native resident of outlying areas; and (II.) who comes from an environment where a language other than English has had a significant impact on the individual's level of English language proficiency; OR (iii.) who is migratory, whose native language is a language other than English, and who comes from an environment where a language other than English is dominant; AND (D.) whose difficulties speaking, reading, writing, or understanding the English language may be sufficient to deny the individual — (i.) the ability to meet the State's proficient level of achievement on State assessments described in section 1111(b)(3); (ii.) the ability to achieve successfully in classrooms where the language of instruction is English; or (iii.) the opportunity to participate fully in society.

SOURCE: P.L. 107-110, Title IX, Part A, Sec. 9101 (25).

is English and to participate fully in society. As discussed below and in Chapter 2 in more depth, this complex definition poses significant problems in measuring the population.

The goals set by the NCLB were designed to ensure that LEP students and immigrant children and youths attain English language proficiency (ELP), and further, that they develop high levels of academic attainment in English and meet the same state academic content and student academic achievement standards as other children (Section 3102(1)). In requiring that all children, including English language learners, reach high standards by demonstrating proficiency in English language arts by 2014, the law challenged the states to develop an integrated system of ELP standards, assessments, and objectives that are linked to states' academic content and student achievement standards set in accordance with other parts of the ESEA.

The part of the legislation that has most changed the landscape is the language that makes it clear that states, districts, schools, and teachers must not only teach ELL students to speak, read, and write English, but they must also hold them to the same high academic standards as all other students. The goal is for all ELL students to demonstrate proficiency in English language arts and mathematics by 2014. Under the ESEA, states now must annually assess ELL students' progress in becoming English language proficient, and they must include these students in annual assessments in all content areas. The states are being held accountable for demonstrating that ELL students are making progress in learning academic subjects. According to the accountability provisions of NCLB, states must include the performance of ELL students in the determination of each school's adequate yearly progress reporting.

The explosive growth in the number of ELL students is another factor in the changed landscape. According to the DoEd, ELL students are the fastest growing educational subgroup in the nation. While the overall school population has grown by less than 3 percent in the last 10 years, the number of LEP students has increased by more than 60 percent in that time (U.S. Department of Education, 2008a, p. 8).

The increased population of ELL students has had a profound influence on the expansion of ELL programs in some states and many localities, putting pressure on states to increase program resources. Between the 2002-2003 and 2007-2008 school years, the period in which data have been collected systematically on LEP students in grades K-12, the count of LEP students increased almost 25 percent, from 3,643,219 to 4,492,068.

In some states, the growth has been profound. For instance, North Carolina and Nevada reported their ELL population growth as 500 and 200 percent, respectively, over the past 10 years (Batlova et al., 2005, as cited in Short and Fitzsimmons, 2007). In California, in 2008, about one-fourth of all students and one-third of elementary school students were English language learners (EdSource, 2008, p.1). This growth has led to a significant increase of programs to support ELL students. The ELL population is quite heterogeneous, and this heterogeneity poses measurement challenges. For example, more than 400 different languages are reported to be spoken by these students, although nearly 80 percent of LEP students speak Spanish (U.S. Department of Education, 2008a, p. vii). Many students come from families that speak multiple primary languages. This heterogeneity poses challenges to the local school systems, generating requirements for special curricula and other instructional resources as well as tailored monitoring, tracking, and assessment. Teaching this heterogeneous student population requires highly qualified teachers with specialized training for teaching such learners, and therefore requires teacher professional development for this task.

The sizable ELL population is a particular challenge because students are at varying levels of ELP and may not be sufficiently proficient in English to demonstrate proficiency in academic content areas. Because they have the task of learning English and academic content simultaneously, it is not surprising that, as a group, they do not meet the proficient level in academic subjects: the academic gap between the group and the non-ELL population is considerable. State data show that the percentage of LEP students who score proficient on a state's language arts and mathematics tests was lower than the state's annual progress goals in nearly two-thirds of the 48 states for which the U.S. Government Accountability Office (2006a, p. 18) obtained data.

Although the NCLB legislation has made a significant contribution to raising awareness about the need to improve ELL students' learning and academic performance, "it has also generated challenges for states to establish a valid accountability system for ELL students" (Wolf et al., 2008, p. 2). NCLB has placed a greater emphasis on addressing the education needs of LEP students than ever before, and Title III was designed specifically to address these needs (U.S. Department of Education, 2005b, p. 1).

Since the passage of the NCLB, additional emphasis has been placed by the DoEd on the development of common core standards. Under the auspices of the National Governors Association's Center for Best Practices (NGA Center) and the Council of Chief State School Officers, a set of state-led common core standards has been developed for English-language arts and mathematics for grades K-12. These standards were developed in collaboration with a variety of stakeholders, including content experts, states, teachers, school administrators and parents.[2]

The adoption of such standards has been promoted by the DoEd's Race to the Top initiative under which consortia of states have been funded to develop an assessment system aligned to the common core standards[3] (U.S. Department of Education, 2010b). This initiative fosters common core standards and is likely to affect programs of education for ELL students, and it may affect state ELP standards and the tests and assessment tools associated with those standards.

ALLOCATING FUNDS FOR TITLE III PROGRAMS

NCLB changed the way in which ELL programs are funded in a very significant and fundamental way. Prior to the Title III amendment, the federal government provided funds for specific projects and services by local educational agencies, but since the implementation of this legislation the funds have been distributed to the state education agencies through a formula grant mechanism. In fiscal 2010, these funds amounted to $750 million: they could grow to $800 million in fiscal 2011 (U.S. Department of Education, 2010b).

The DoEd allocates the Title III funds through the following formula, after funds are reserved for discretionary grant awards and other purposes as specified in the legislation: 80 percent of the remaining funds are allocated to states on the basis of each state's share of the national estimate of LEP students, and 20 percent are allocated on the basis of each state's share of the national estimate of immigrant children and youth (see Box 1-2).

There is a minimum state allocation of $500,000, and states are required to use up to 15 percent of their allotments for school districts with significant increases in school enrollment of immigrant children and youth. See Box 1-3 for the legislative language. The states in turn allocate the federal funds to local education agencies (LEAs) on the basis of the number of LEP students served in those LEAs.

---

[2]For information on these standards, see http://www.corestandards.org/about-the-standards [December 2010].

[3]See the department's announcement of winners at http://www.ed.gov/news/press-releases/us-secretary-education-duncan-announces-winners-competition-improve-student-asse [December 2010].

BOX 1-2
State Allotments

Allotments for use in the department's formula allocations are specified in Section 3111(c) (3) of Elementary and Secondary Education Act, Title III, Part A, as amended by the No Child Left Behind Act of 2001:

(3) STATE ALLOTMENTS—
(A) IN GENERAL—Except as provided in subparagraph (B), from the amount appropriated under section 3001(a) for each fiscal year that remains after making the reservations under paragraph (1), the Secretary shall allot to each State educational agency having a plan approved under section 3113(c) — (i) an amount that bears the same relationship to 80 percent of the remainder as the number of limited English proficient children in the State bears to the number of such children in all States; and(ii) an amount that bears the same relationship to 20 percent of the remainder as the number of immigrant children and youth in the State bears to the number of such children and youth in all States.
(B) MINIMUM ALLOTMENTS—No State educational agency shall receive an allotment under this paragraph that is less than $500,000.

Allowable Data Sources

The Title III legislative mandate left it to the DoEd to determine the source of information to be used to determine the number of LEP and immigrant students to be used in the allocation formula, but it stipulated only two allowable data sources: (1) estimates of the population to be served from the U.S. Census Bureau's American Community Survey (ACS) or (2) counts of the number of students being assessed for ELP by the states.

ACS Estimates

Since 2005, the department has been using only the ACS for estimates of two groups, the LEP and immigrant populations. ACS estimates are based on a nationwide household survey conducted by the U.S. Census Bureau (for a description and analysis, see National Research Council, 2007). The survey provides information on the U.S. population at the national, state, county, city, and neighborhood levels and for specific demographic groups, including racial and ethnic groups and children. For the LEP population component of the formula, the estimates used in the allocation formula are based on the responses to questions about the English-speaking ability of school-aged household members as a proxy for the number of LEP children

BOX 1-3
Legislative Mandate for Estimating the Number of LEP Students

The mandate for estimating state numbers of LEP students for use in the department's formula allocations is found at Section 3111(c) (4) of Elementary and Secondary Education Act, Title III, Part A as amended by the No Child Left Behind Act of 2001:

(4) USE OF DATA FOR DETERMINATIONS—
(A) IN GENERAL—In making State allotments under paragraph (3), for the purpose of determining the number of limited English proficient children in a State and in all States, and the number of immigrant children and youth in a State and in all States, for each fiscal year, the Secretary shall use data that will yield the most accurate, up-to-date numbers of such children and youth.
(B) SPECIAL RULE—
(i) FIRST 2 YEARS—In making determinations under subparagraph (A) for the 2 fiscal years following the date of enactment of the No Child Left Behind Act of 2001, the Secretary shall determine the number of limited English proficient children in a State and in all States, and the number of immigrant children and youth in a State and in all States, using data available from the Bureau of Census or submitted by the States to the Secretary.
(ii) SUBSEQUENT YEARS—For subsequent fiscal years, the Secretary shall determine the number of limited English proficient children in a State and in all States, and the number of immigrant children and youth in a State and in all States, using the more accurate of—
(I) the data available from the American Community Survey available from the Department of Commerce; or
(II) the number of children being assessed for English proficiency in a State as required under section 1111(b)(7).

SOURCE: Section 3111(c) (3) of Elementary and Secondary Education Act, Title III, Part A, as amended by the No Child Left Behind Act of 2001.

in the state. For the immigrant component of the formula, the estimates are based on responses to questions on place of birth and year of immigration.[4]

For the LEP component, two ACS questions are used: "Is a language other than English spoken in the home?" and if the response is yes, "How well does household member X speak English?" Four choices are given for the second question: "Very well," Well," "Not well," and "Not at all." The LEP estimates that are reported to

[4]Before 2005, the LEP estimates were based on similar questions in the 2000 census long-form sample, and the immigrant estimates were based on state counts of recent immigrant students enrolled in grades K-12 in public and private schools.

the department from the ACS represent the total number of persons aged 5-21 for whom the answer is anything less than "Very well."

Details of the overall count are available by cross-tabulating the data with responses to other questions. For example, it is possible to differentiate between students who attend public and those who attend private schools and to present the data by different age cohorts. In this report, all of the data from the ACS that pertain to the ELL population start at age 5 because the language ability question is only asked about 5 years and older. Data are presented for both the total number of people aged 5-21 and those aged 5-18 because the latter group better represents the elementary and secondary school age population (in Chapters 2 and 5). Similarly, data are presented for the total population and also only for those enrolled in public schools in order to facilitate comparisons with the state counts, which represent ELL students in public schools (in Chapters 2 and 5).[5] A summary of these variables as used in the legislative mandate, the current ACS data for allocation purposes, the state counts, and the ACS data used in this report for comparisons with the state data are shown in Table 1-1.

For the immigrant component, two other ACS questions are used: "Where was household member X born?," and if the response is "born abroad," "When did household member X come to the United States?" The recent immigrant estimates that are reported to the DoEd from the ACS represent the number of persons aged 3-21 who were born abroad and arrived in the United States no more than 3 years prior to the survey.

Several survey methodology factors affect the accuracy and precision of the ACS estimates, especially the sample design, mode of interviewing, and selection of the respondent.

- *Sample Design and Size* Each year, ACS questionnaires are sent to 3 million household addresses, and about 2 million responses are ultimately recorded in ACS data files. The responses are given unequal weighting due to subsampling of households that do not respond by mail or telephone, which increases variability of the sample weights and therefore the imprecision of the estimates relative to an equal probability design. Because the ACS estimates reflect relatively small sample sizes for a single year, the estimated numbers of LEP and immigrant children have varied significantly from year to year. Consequently, the relative allocations of funding across the states have also varied significantly from year to year.
- *Mode of Interviewing* About 50 percent of ACS responses come from mailed-back questionnaires; another 8-9 percent from computer-assisted telephone interviewing; and the final 40-42 percent from computer-assisted personal interviewing of about one-third of the households that did not respond by

[5]It is not possible for the DoEd to replicate the legal definition of the LEP population (aged 3-21) presented in Box 1-1 because ACS data are not available for those under 5 years of age.

TABLE 1-1 Summary Definitions of Selected Variables Used in This Report

Variable	Legislative Mandate	ACS Estimate for Allocation Purposes	State Counts	ACS Estimate for Use in Comparisons with State Counts
Age	3-21 years	5-21 years	5-18 years	5-18 years
School enrollment status	Enrolled or preparing to enroll in a elementary and secondary school	Enrollment status not specified	Enrolled in public schools	Enrolled in public schools in the last 3 months
English speaking ability	Those "whose difficulties speaking, reading, writing, or understanding the English language may be sufficient to deny the individual — (i.) the ability to meet the State's proficient level of achievement on State assessments; (ii.) the ability to achieve successfully in classrooms where the language of instruction is English; or (iii.) the opportunity to participate fully in society"	Speaks English "less than very well"	An unduplicated count of all students in the state who meet the definition of LEP, which includes newly enrolled students whether or not they receive Title III services	Speaks English "less than very well"

mail or telephone. These different modes of response may affect the comparability of the responses, and because LEP and immigrant children are more likely to be in households that respond by mail versus those who are in households that respond through an interaction with an interviewer, the responses may be biased.

- *Choice of Respondent* One person in a household typically provides responses for all household members and that person's judgment of the young family member's English speaking ability (as solicited in the wording of the question) may bias reporting of the English speaking ability for children and youth or of the reporting of year of immigration. Household responses may also differ from results that would be obtained in other ways—for example, from state tests or records—in ways that could bias the ACS estimates.

State Estimates

State-provided counts of LEP children basically reflect the number of students in public schools (including charter schools) who are assessed for proficiency in English in a given year. Although states use a variety of instruments and procedures to identify which students are tested, most start with a questionnaire to identify students who live in homes in which a language other than English is spoken. Students in these households are then assessed for English proficiency using state- or district-approved instruments and procedures. Students who fail to demonstrate proficiency on these assessments are designated as LEP, deemed eligible to receive services under ESEA Title III, and entered into the count that a state reports to the DoEd.

Many factors may affect the counts of LEP students estimated by the states, given the U.S. educational system's tradition of state and local control of education, including authority for determining the state's definition of LEP, the criteria used to classify students as LEP or not LEP, the test or tests used to assess English language proficiency, and the criteria that the state uses to determine when LEP students are deemed ready to exit LEP status. There are many sources of differences. Some states allow local school districts to determine the procedures, tests, and criteria for identifying and classifying students as LEP, which results in counts that are derived in different ways even within a state. States may change their procedures, tests, and criteria over time, which results in counts that are derived in different ways over time.

These differences among states in their procedures, tests, and criteria result in counts that are derived in different ways from state to state. Yet a high degree of comparability within states and among states is essential for equitable distribution of funds in a formula allocation that is based on shares of a fixed appropriated amount. And a high degree of comparability across time is essential to ensure that year-to-year changes in allocations reflect actual changes in LEP students and not changes in the procedures for testing and identifying students as LEP.

REVIEW BY THE U.S. GOVERNMENT ACCOUNTABILITY OFFICE

Given the differences in estimation practices summarized above, it is not surprising that the two allowable data sources used in the computation of the distribution of Title III funds have tended to yield marked differences in the amounts that would be allocated to some states. The use of the ACS estimates has also led to fluctuation between years in the funding amounts provided to states. As a result, at the request of Congress the U.S. Government Accountability Office (GAO) carried out a comprehensive review of the data sources in 2005 (U.S. Government Accountability Office, 2006b, p. 2). The GAO review compared the dollar amounts that would be allocated for LEP students by using state testing results and by using the ACS sample estimates based on responses to the subjective English ability questions.

In a simulation with data on 12 states, the GAO confirmed that the differences in methodology yielded very different LEP estimates, which could result in very different fund allocations to the states. The review showed that ACS estimates were higher than counts based on state data in six cases and lower in six cases; the differences were sometimes quite striking. For example, based on data for the 2004-2005 school year, the ACS estimate of LEP students in California was almost 50 percent lower than the state's estimate, and the ACS estimate for New York was almost 40 percent higher than the state's estimate.

These widely different results, coupled with fundamental differences in how the numbers are derived, led the GAO to conclude that "ACS and State data each measure different populations in distinct ways and it is unclear how well either of the two data sources captures the population of children with limited English proficiency" (U.S. Government Accountability Office, 2006b, p. 3). GAO recommended that the DoEd provide clear instructions to states on how to provide data specified in the ESEA on the number of LEP students. GAO also recommended that the DoEd develop and implement a methodology for determining which is the more accurate of the two sources of data and seek authority to use statistical methodologies to reduce the variability associated with ACS data.

In its evaluation of the state counts, the GAO study found other problems (U.S. Government Accountability Office, 2006b, p. 23):

> With regard to data states collect on the number of children and youth who are recent immigrants, state officials expressed a lack of confidence in these data. State officials in some of the 12 study states told us that these data were not very reliable because school and school district officials did not ask about immigration status directly. Some state and school district officials told us that in order to determine whether a student should be classified as a recent immigrant, they relied on information such as place of birth and the student's date of entry into the school system. Officials in one state told us that in the absence of prior school documentation, they made the assumption that if a student was born outside the U.S. and entered the state's school system within the last 3 years, then the student was a recent immigrant.

In a presentation to our panel, the authors of the GAO review reiterated the report findings that state data were incomplete, inconsistent, and of poor quality in the early years of the program and that these deficiencies could affect the distribution of Title III funds. They also highlighted GAO findings that state counts from the ACS showed substantial variation for many states in the early years of the ACS, when sample sizes were much smaller than they are now, and that the variability significantly affected the amounts allocated to the states.

In a response to the GAO report that was contained in the report, the DoEd agreed with the GAO findings, but argued that ACS data were selected as the source for its allocations because of the problems with the state administrative data sources (U.S. Government Accountability Office, 2006b, p. 49). The GAO report did document some of these problems with the state data, which seemed to be related to federal requests for the number of LEP students assessed for English proficiency each year. One problem was that the instructions did not include clear definitions (U.S. Government Accountability Office, 2006b, p. 12): "It was unclear whether states should provide the number of students screened for English proficiency, the number of students who were already identified as [LEP] who were then assessed for their proficiency or a combination of the two numbers."

Another inconsistency reflected lack of clarity about whether states were to provide an unduplicated count of students or not. For example, some states use more than one assessment to evaluate a student's English proficiency (such as separate assessments to assess skills in reading, writing, speaking, or listening); in these states, students could be reported more than once. According to the GAO (U.S. Government Accountability Office, 2006b, p. 12): "As a result, some states included duplicate counts of students, and in other states, these data included other student counts (based on screening of new students rather than assessments of already identified students)."

In its response to the GAO report, the DoEd did not consider developing a methodology to compare the relative accuracy of the two approved data sources because of the serious issues with the state-provided data, but they pledged to revisit the GAO recommendation in the future as the quality of state data improved (U.S. Government Accountability Office, 2006b, pp. 48-50). The department did develop a plan to improve the quality of the data collected from the states. This plan included revising the instructions on the Consolidated State Performance Reports, comparing recent data to data for prior years, and incorporating data edits and checks to guide state officials when they entered data electronically. Department officials expected that these changes would improve data quality, beginning with the 2005-2006 school year. The planned changes were made, and the state data appear to be more complete with fewer year-to-year fluctuations (see details in Chapter 4). Thus, it is now appropriate to compare the relative accuracy of the data sources and to assess whether state data are appropriate for funding purposes.

THIS STUDY AND THIS REPORT

Against this backdrop, the DoEd asked the National Research Council's Committee on National Statistics and Board on Testing and Assessment to convene a group of experts to review and make recommendations regarding the two allowable data sources to use for future Title III formula allocations, for both LEP and recent immigrant students. In evaluating the two sources, the Panel to Review Alternative Data Sources for the Limited-English Proficiency Allocation Formula under Title III, Part A, Elementary and Secondary Education Act, was asked

> to review alternative data sources for use in formula grants to states to ensure that limited English proficiency (LEP) children and youth attain English language proficiency under Title III, Part A, of the Elementary and Secondary Education Act. The formula includes two components: LEP children and youth and recent immigrant children and youth. The panel will evaluate the two currently allowable sources of estimates of each component—those from the American Community Survey (ACS) conducted by the U.S. Census Bureau and those from the results of state tests on English proficiency or state records of immigrants. In evaluating the two sources for each component, the panel will consider the accuracy and precision of estimates derived from each; what methodological, demographic, and other factors influence the estimates from each source; and what statistical or data collection methods might

> increase the accuracy and precision of the estimates used in Title III allocations. In evaluating the ACS data, the panel will review not only the questions that are used to indicate limited-English proficiency and immigrant status (in particular, their accuracy), but also the literature to determine the experience of other household surveys in reporting of these items that may suggest research and development to improve their reporting on the ACS. In evaluating state tests of English proficiency, the panel will make a comprehensive assessment of the quality and comparability of such tests among states. Finally, the panel will determine if there are other data sources or methods that might be preferable to the two permitted by statute. On the basis of its information-gathering activities, the panel will deliberate, make recommendations, and publish these recommendations along with supporting findings as an independent NRC report at the conclusion of its study.

In addition to addressing these questions, the panel probed deeply into the quality, comparability, and usefulness of state tests of ELP. The panel also reviewed the literature on the accuracy of household's reports of LEP and recent immigrant status in other surveys: that research may provide insights for ACS reporting and the kinds of methodological research and development that the Census Bureau could conduct to improve the ACS estimates.

Study Data and Information

The panel began its work by reviewing the work of the GAO in its 2006 report. In addition, for the ACS, the panel determined the availability of information about the accuracy and precision of responses to the questions on English speaking ability that are used to estimate LEP children aged 5-21 by state, as well as to the questions on place of birth and year of immigration that are used to estimate children aged 3-21 by state who moved to the United States within 3 years of the survey date. With regard to precision, the panel analyzed the effects on the use of the ACS estimates for allocation purposes by combining estimates across more than 1 year in order to reduce sampling error. With regard to nonsampling error, the panel used the ACS public-use microdata sample files to evaluate patterns of nonresponse and imputation to the English speaking questions.

To facilitate its work with the ACS, the panel benefitted from the willingness of the DoEd to share the tabulations that the Census Bureau had done for allocation purposes. The department also facilitated the preparation of special 3-year ACS tabulations so the panel could assess the appropriateness of these data for allocation purposes.

The panel extended its investigation into likely nonsampling errors in ACS estimates of LEP and recent immigrant students by conducting a broad review of the survey research literature on the validity of household reports of both ELP and immigrant status. The panel also reviewed prior studies of the questions in the long forms of the 1990 and 2000 censuses and the precursor survey to the ACS in order to consider recommendations on the kinds of research the DoEd should request the Census Bureau carry out to evaluate the LEP and immigration questions more precisely and, ultimately, to improve them. For states' estimates of their numbers of LEP students, the panel examined the published reports that document the procedures used by the states to identify and classify students as LEP and the assessments used to evaluate their ELP. The panel also heard presentations from a sample of Title III directors on the procedures, tests, and criteria they use, as well as from a sample of the organizations that develop ELP tests.

Features of the state practices and tests that were examined by the panel included definitions of LEP; intake processes, including home language surveys, other assessments, and teachers' observations; cutoff scores used to determine English language proficiency on these tests; the types of ELP tests that are used and their mode of administration; and provisions for "exiting" students from LEP status. Moreover, the panel's review of the ELP tests extended beyond the yes-no question of whether state assessments of LEP students have become sufficiently standardized to justify their consideration for Title III allocations. The panel gathered more in-depth information on the ELP assessments in several states in order to provide the basis for findings and recommendations about ways in which the tests could affect the relative counts of ELL students in the states. However, this in-depth review, which included a polling of the states to ensure that the data were up to date, did not include a comparative review of the actual content of the various proficiency assessments used by the states. Such a review would have been beyond the scope of the panel's charge and require resources beyond those available to the panel.

The panel also reviewed issues concerning counts of recent immigrant students. It looked at the procedures for those counts in order to understand the comparability and quality of state reports of recently immigrated children and youth.

Early on in its deliberations, the panel decided to limit its focus to the two eligible sources of data for the Title III allocations. There are no other household-based surveys that have both a sufficient sample size and the type of questions that permit identification of an ELL population with acceptable precision at the state level, nor are there any administrative record sources that are as germane to the allocations as the state counts of ELL students.

If the panel were to have judged the two allowable data sources to be unsatisfactory for the purpose of allocating federal funds at the state level, it may have been advantageous to consider model-based estimates, along the line of the Small Area Income and Poverty Estimates (SAIPE) Program of the Census Bureau to provide more current estimates of selected income and poverty statistics than those from the most recent decennial census for school districts, counties, and states.[6] However, the two allowable data sources each meet the basic criteria for service as a basis of allocations, so the creation of a model-based estimate is not required.

---

[6]The SAIPE Program provides updated estimates of income and poverty statistics for the administration of federal programs and the allocation of federal funds to local jurisdictions under programs such as Title I of the ESEA. These estimates combine data from administrative records, intercensal population estimates, and the decennial census with direct estimates from the ACS to provide consistent and reliable 1-year estimates.

If the panel's recommendation that an eventual allocation formula should be based on both the ACS and the state counts is deemed unsatisfactory, the DoEd might decide to devote resources to develop a model that combines the ACS estimates and the state counts. In the judgment of the panel, however, neither of the two allowable sources has flaws that require consideration of alternative data sources or methods at this time.

Definition of an LEP Student

In approaching its charge, the panel was mindful that, unlike the case with many other pieces of legislation that prescribe a formula for allocation of federal funds to other units of government, the law establishing the Title III allocations was relatively specific in defining the *allowable data sources* for formula elements. However, the law failed to define the *specific data elements* that should be drawn from these data sources for the computation of the allocation formula.

The ACS definition of an LEP student can only be a proxy for the official LEP definition (see above). Though the ACS collects objective demographic and immigrant information and subjective information on English speaking ability, it collects no information on the ability of a student to meet the state's proficient level of achievement on state assessments, nor does it directly measure command of the English language for classroom success or full participation in society.

A measure based on the ACS definition implicitly assumes that, at some level of reported English speaking ability, a student will encounter difficulty in meeting the state's proficient level of achievement on state assessments including the English language arts assessment and assessments of reading, mathematics, and science, or have difficulty learning content in English, or have difficulty participating fully in society.

A definition based on state administrative records would not be similarly encumbered with the need to make assumptions about the relationship between reported English speaking ability and the ability to be proficient and to be successful in learning content in English in the classroom. For the most part, these criteria are measurable through tests and direct classroom observation and are readily summarized by means of such measures as ELP test scores and achievement data. However, the "ability to fully participate in society" is a subjective criterion.

The official definition of LEP student that is promulgated by the department for use in state reports is the legal definition (U.S. Department of Education, 2010a, sec. 4.3, p. 12). This definition is carried forward into the Consolidated State Performance Reporting (CSPR) system in which the data are reported (as CPSR item number 1.6.2.1), and they are entered into the DoEd's official EDFacts database (as data group 678 with file specification 141). This precise definitional trail has now been consistent for two reporting periods, 2007-2008 and 2008-2009. Thus, in terms of adherence to the letter of the law, state data are now in compliance.

Unfortunately, definitional consistency throughout the reporting chain is not a sufficient basis on which to judge that a state administrative data system yields a "more accurate" estimate of the number of LEP students, as is required by ESEA. To be more accurate, the statutory official definition would have to apply a consistent basis to ensure uniform measurement within and across states.

A good and usable definition is not only consistent, it is also a transparent offshoot of the operation of the programs within the states and stable over time. The task of arriving at one consistent definition applicable across the states is particularly difficult because of the rich variety of programs and measures used by the states and localities to meet their obligations under the law (and its interpretations), as well as the many changes in state practices and reporting procedures over time, particularly those having to do with students' entry and exit from the program. In regard to defining LEP, the department has concluded that there is "no one, common, approved method to operationalize the term, either for initial identification purposes or for ultimate exit from an Language Instruction Educational Program (LIEP) or the LEP category" (U.S. Department of Education, 2008a, p. 7). The 2006 GAO study documented no less than three operational definitions that could be employed to identify the LEP population: see Table 1-2.

Since 2006, considerable progress has been made by the department and the state education agencies in refining the data collected on the LEP population in the CSPR system. New policies about how data are to be collected and aggregated have also emerged from this effort. Thus, for purposes of this report, the definition that has been selected for the analysis is the definition used in collecting operational data from state agencies, which are the data as reported in the Consolidated State Performance Reports as Code 1.6.2.1: the unduplicated number of limited English proficient students enrolled in an elementary or secondary school at any time during

TABLE 1-2 Operational Definitions of the ELL Population

Definition	Purpose	How Measured
The number of students with limited English proficiency in grades K-12 who are assessed for English proficiency	ESEA allowable data for Title III allocation	States develop assessment instruments and practices, with data collected by state education agencies.
The number of students identified as limited English proficiency in grades K-12	State standards for identification of the population needing services	States use various methods of identifying the population, including home language surveys or teacher observation reports, which are administered by local education agencies with data collected by state agencies.
The number of students enrolled in state and local Title III programs	Administrative counts of program participants	State education agencies collect this data from local education agencies as an administrative requirement.

SOURCE: U.S. Government Accountability Office (2006b, p. 14).

the school year. The panel selected this measure primarily because it is an inclusive number, pertaining to the total LEP population not just those who have been assessed under provisions of the NCLB. Moreover, this definition is expedient. In the department's reporting scheme, it is one of the few measures for which there is a comparable historical time series. The measure also relates directly to subordinate measures that have also been consistently collected over time, such as grade level, home language, and language proficiency level. The definition of the data items from the ACS is the same as that used in the GAO report. This definition derives from the responses to the ACS survey question on the number of persons aged 5 to 21 who speak a language other than English at home and report speaking English less than "very well."

Overview of the Report

In the ensuing chapters, the panel first discusses the desired characteristics of allocation formulas and then assesses the two allowable sources in terms of their relative ability to fulfill those desired characteristics.

Chapter 2 assesses the ACS. It provides a summary of the survey and how the ACS estimates are presently used to make Title III allocations to states. It evaluates the quality of those estimates in terms of sampling properties, precision, sensitivity, coverage, and consistency.

Chapter 3 discusses the ELP assessments used by the states, describes their features, and examines the ways in which they differ. It considers the technical quality of these tests and focuses on the extent to which they are likely to yield valid and comparable decisions across the states.

Chapter 4 focuses on state policies and procedures for initially identifying ELL students, measuring their progress in becoming English proficient, and determining when they are ready to be reclassified as former ELL students (and exited from programs for English as a second language).

Chapter 5 discusses the comparability of the estimates of the ELL population derived from the ACS and the state administrative record counts.

Chapter 6 discusses the comparability of the estimates of the immigrant student population from the ACS and those reported to the states by local education agencies.

The concluding chapter considers possible decision criteria for making the choice between the two allowable data sources, rates the sources by these criteria, and presents the panel's recommendations.

2

American Community Survey Estimates

This chapter provides background information on the American Community Survey (ACS) estimates of the number of English language learner (ELL) students that are used for computing each state's share of the national estimate for the allocation of Title III funds. The chapter first provides a summary of the ACS and then assesses the evidence on the quality of the ACS estimates. The third section presents the ACS estimates, and the last section describes the properties of the estimates in terms of their sampling properties, precision, consistency, sensitivity, and coverage.

THE AMERICAN COMMUNITY SURVEY

Characteristics

Although the ACS is a new survey—its first products were released in 2006, after a decade of testing and development by the Census Bureau—it is a very important one. Unlike the long-form sample of the decennial census, which it replaced, it is a significant ongoing undertaking that covers some 2 million households each year. It provides the capacity for the Census Bureau to produce estimates for 1-year, 3-year, and 5-year periods and for successively broader tabulation coverage of geographic areas.

Other characteristics of the ACS enhance its value to users (National Research Council, 2007, p. 2), especially in comparison with the census long form: it is timely, with data products introduced just 8-10 months after collection; frequent, with products updated each year; and of relatively high quality, as measured by the completeness of response to survey questions. Given these characteristics, a great number of uses have already been implemented, and many more have been identified for the ACS data, including the allocation of federal funds for programs that support activities in states and localities. A recent study by the Brookings Institution found that, in fiscal 2008, "184 federal domestic assistance programs used ACS-related datasets to help guide the distribution of $416 billion, 29 percent of all federal assistance. ACS-guided grants accounted for $389.2 billion, 69 percent of all federal grant funding" (Reamer, 2010, p. 1).

However, some characteristics of the ACS limit its usefulness for particular applications or levels of detail. Like the census long form, the ACS is a sample survey. Even with the aggregation of data for 5-year estimates, the ACS sample is significantly smaller than the census long-form sample it replaced, and it therefore has considerably larger margins of error in the sample estimates. In addition to smaller sample size, the ACS sample has greater variation because of greater variation in sample weights because of the subsampling of households for field interviews from among those that do not respond to the mail or telephone contacts. Some uncertainty in the ACS estimates is also introduced by the use of postcensal population and housing estimates as controls for the survey over the course of the decade. These estimates are applied at a less detailed level than census controls, and they are indirect estimates rather than a product of a simultaneous census activity (as were the census controls for the long-form sample). However, some of the characteristics of the ACS mitigate these negative aspects. Because of extensive follow-up, the response rates are higher than response rates achieved with the census long form, and because a higher proportion of ACS responses are through the intervention of an interviewer, the overall quality of the responses tends to be higher.

The effects of the larger sampling errors fall most heavily on the data for small areas and small population subgroups. Later this is illustrated in Table 2-2, which shows that standard errors are proportionally largest for the smallest states with regard to the critical data element used in the allocation of Title III funds. The relative lack of precision for smaller states suggests the need to accumulate data for 3-year and 5-year periods, rather than using 1-year estimates, in order to achieve sufficient precision for some data elements, such as English speaking ability. The issues attending the selection of the appropriate ACS period are extensively discussed below.

Background

It is useful to trace some of the significant events in the evolution of the ACS in order to understand the environment that led to tradeoffs that, in turn, set the objectives for this new survey. After the 1990 census, there were growing concerns, shared by some members of Congress, that the long-form questionnaire had response issues that marginalized its utility. In that census, 29 percent of the households that received the long form failed to mail it back, compared with 24 percent of households that received the short form (National Research Council, 2004, p. 100). Some observers thought that this differential contributed to the poorer coverage of the population in 1990 in comparison with 1980. At the same time, there was increasing interest in obtaining more frequent population estimates for small areas.

To counter this problem of declining long-form response rates and to provide more frequent data for small areas, in 1994 the Census Bureau decided to move toward a continuous measurement design similar to one that had been proposed years earlier by Leslie Kish (see National Research Council, 1995, p. 71). This continuous measurement survey was named the American Community Survey, and the Census Bureau set a goal of conducting a short-form-only census in 2010 and to fully implement the ACS by then. It was expected that the ACS could provide estimates for small areas that were about as precise as long-form-sample estimates for small areas by accumulating samples over 5 years. However, very early in the development process, rising costs led to a decision to scale back the originally planned size of 500,000 housing units per month to a sample of 250,000 housing units per month (National Research Council, 1995, p. 127). This decision to reduce the desired sample size had a significant deleterious effect on the ability of the ACS to provide reliable 1-year data for small areas.

Design

Data Collection

Each month, the ACS questionnaire—which is similar in content to the old census long form—is mailed to 250,000 housing units across the nation. The units have been sampled from the Census Bureau's Master Address File using a probability sample design in which housing units in small areas are oversampled. As with the long form of the census, response to the ACS is required by law.

The ACS mail questionnaire uses a matrix layout for questions on sex, age, race, ethnicity, and household relationship. It provides space for information on five household members; information on additional household members is gathered through a follow-up telephone survey. The ACS instructs the household respondent to provide data on all people who, at the time of completing the questionnaire, have been living or staying at the household address for more than 2 months (including usual residents who are away for less than 2 months). Individuals in the ACS samples that reside in group quarters (such as college dormitories and prisons) are counted at the group quarters location, in effect applying a de facto residence rule regardless of how long an individual has lived or expects to live in the group quarters.

The residential housing unit addresses in the ACS sample with usable mailing addresses—about 95 percent of each month's sample of 250,000 addresses—are sent a notification letter 4 days before they receive a questionnaire booklet, and a reminder postcard is sent 3 days after the questionnaire mailing. Whenever a questionnaire is not returned by mail within 3 weeks, a second questionnaire is mailed to the address. If there is no response to the second mailing, and if the Census Bureau is able to obtain a telephone number for the address, trained interviewers conduct telephone interviews using computer-assisted telephone interviewing (CATI) software.

The CATI operation benefits from several quality assurance programs. The software prevents common errors, such as out-of-range responses or skipped questions. Full-time call center staff are carefully trained and provided with periodic training updates. New interviewers receive standard CATI training and a workshop to specifically train them on how to handle refusals. New interviewers are monitored regularly and even qualified interviewers are monitored periodically to make sure they continue conducting interviews in a satisfactory manner. In addition, Census Bureau supervisors at the call centers monitor interviewers' work to check for other errors, such as keying a different answer from the one the respondent provided or failing to follow procedures for asking questions or probing respondents for answers to questions. The Census Bureau has found its monitoring to be effective in controlling telephone interviewer errors. Consequently, the ACS, using CATI instruments and procedures, is more accurate than the census long form in that it obtains more complete information than was obtained on the Census long form (National Research Council, 2007, p. 161).

Interviewers also follow up on a sample of households: those for which no mail or CATI responses have been obtained after 2 months, those for which the postal service returned the questionnaire because it could not be delivered as addressed, and those for which a questionnaire could not be sent because the address was not in the proper street name and number format. The follow-up is in person for 80 percent of the housing units and by telephone for 20 percent. For the in-person interviews, the data are collected though computer-assisted personal interviewing (CAPI).

For cost reasons, the personal interview follow-up is conducted on a sample basis: it includes about two-thirds of unusable addresses and between one-third and one-half of usable addresses in each census tract, depending on the expected mail-back and CATI response rate for the census tract. Interviewers also visit group quarters in person to collect data from residents, using paper-and-pencil questionnaires.

Since it is considered a part of the decennial census, the ACS collects data under legal protections[1] with confidentiality requirements. Following the law, the Census Bureau pledges to respondents that their responses will be used only for statistical purposes and not for any kind of administrative or enforcement activity that affects the household members as individuals. This confidentiality protection is one reason for the high response rate to the ACS, even on somewhat sensitive topics.

Because ACS data are collected on an on-going basis, data products are available each year and do not pertain to a specific point in time. The 1-year estimates operate on 12 months of data collected during the preceding calendar year. The 3-year estimates are produced using 36 months' worth of responses, and the 5-year estimates are produced from 60 months' worth of responses. For the range of sample sizes used in producing ACS estimates for each state, see Table 2-1.

The data used to generate the period estimates include all of the mailed back, CATI, and CAPI responses (including additional information obtained by telephone

[1]Data Protection and Privacy Policy, available: http://www.census.gov/privacy/data_protection/federal_law.html [May 2010].

for incompletely filled out mail questionnaires). The major data processing steps are coding, editing, and imputation; weighting; and tabulation.

Coding, Editing, and Imputation

The first data processing step for the ACS is to assign codes for write-in responses for such items as ancestry, industry, and occupation, which is done with automated and clerical coding procedures. Then the raw data, with the codes assigned to write-in items and various operational data for the responses, are assembled into an "edit-input file." Computer programs review the records on this file for each household to determine if the data are sufficiently complete to be accepted for further processing and to determine the best set of records to use in instances when more than one questionnaire was obtained for a household. Computer programs then edit the data on the accepted, unduplicated records in various ways. Computer programs also supply values for any missing information that remains after editing, using data from neighboring households with similar characteristics. The goal of editing and imputation is to make the ACS housing and person records complete for all persons and households.

Weighting

The weighting process is designed to produce estimates of people and housing units that are as complete as possible and that take into account the various aspects of the complex ACS design. The edited, filled-in data records are weighted in a series of steps to produce period estimates that represent the entire population.

The basic estimation approach is a series of steps that accounts for the housing units probability of selection, adjusts for nonresponse, and applies a ratio estimation procedure that results in the assignment of two sets of weights: a weight to each sample person record (both household and group quarters persons) and a weight to each sample housing unit record. Ratio estimation takes advantage of auxiliary information (population estimates by sex, age, race, and Hispanic origin, and estimates by total housing units) to increase the precision of the estimates, as well as to correct for differential coverage by geography and demographic detail. This method also produces ACS estimates consistent with the estimates of population characteristics from the Population Estimates Program of the Census Bureau and the estimates of total number of housing units for each county in the United States.

Tabulations and Data Releases

The final data processing steps are to generate tabulations, profiles, and other data products, such as public-use microdata samples (PUMS). Beginning in summer 2006, the Census Bureau began releasing 1-year estimates from the previous year for

TABLE 2-1 ACS Sample Sizes: Initial Addresses and Final Interviews, by Type of Unit

	ACS 2005		ACS 2006			
	Housing Units		Housing Units		Group Quarters	
State	Initial Addresses Selected	Final Interview	Initial Addresses Selected	Final Interview	Initial Sample Selected	Final Interview
Alabama	51,050	31,274	51,063	32,647	2,767	1,997
Alaska	9,740	5,759	9,739	5,835	485	337
Arizona	51,685	32,749	52,511	33,718	2,609	1,971
Arkansas	32,648	20,052	32,608	20,825	1,873	1,567
California	266,324	172,287	265,521	178,666	19,583	14,783
Colorado	45,086	29,612	45,053	30,623	2,523	1,974
Connecticut	28,885	20,652	28,651	21,357	2,651	2,266
Delaware	9,722	6,208	9,951	6,411	557	467
District of Columbia	5,941	3,684	5,884	3,672	889	587
Florida	157,536	99,565	159,011	103,089	9,256	6,894
Georgia	77,261	47,171	78,573	49,925	5,805	4,269
Hawaii	12,295	7,627	12,054	7,629	833	598
Idaho	15,165	9,953	15,070	10,378	785	476
Illinois	118,210	80,473	117,521	82,815	7,692	6,076
Indiana	60,872	42,812	60,382	43,302	4,355	3,520
Iowa	38,852	28,729	38,680	29,264	2,592	2,034
Kansas	32,644	22,391	32,338	23,097	2,022	1,580
Kentucky	41,734	27,883	41,834	28,658	2,916	2,214
Louisiana	46,953	27,324	46,815	28,573	3,349	2,487
Maine	24,443	14,842	24,167	15,954	865	582
Maryland	45,975	31,474	45,698	32,435	3,266	2,467
Massachusetts	53,543	37,037	52,988	37,990	5,374	3,950
Michigan	123,933	85,771	123,111	88,400	5,817	4,287
Minnesota	77,962	55,645	77,828	57,762	3,313	2,634
Mississippi	28,396	16,177	28,350	16,829	2,407	1,652
Missouri	64,438	43,493	64,434	44,640	3,962	3,241
Montana	14,248	9,076	14,302	9,482	601	478
Nebraska	25,458	18,002	25,254	18,307	1,252	1,036
Nevada	20,360	12,660	21,334	13,498	815	686
New Hampshire	14,933	9,877	15,078	10,352	858	662
New Jersey	72,896	49,132	72,297	50,641	4,802	3,783
New Mexico	19,901	11,862	19,895	12,397	897	674
New York	183,793	116,910	181,711	121,011	14,249	11,484
North Carolina	83,176	53,038	84,642	55,417	6,225	4,592
North Dakota	11,643	8,066	11,622	8,258	592	502
Ohio	110,366	78,913	109,651	80,011	7,341	5,852
Oklahoma	46,827	28,358	46,478	29,492	2,691	2,184
Oregon	33,884	23,379	33,893	23,785	1,873	1,347

ACS 2007				ACS 2008			
Housing Units		Group Quarters		Housing Units		Group Quarters	
Initial Addresses Selected	Final Interview	Initial Sample Selected	Final Interview	Initial Addresses Selected	Final Interview	Initial Sample Selected	Final Interview
51,179	32,345	2,699	1,999	51,817	31,973	2,533	2,109
9,751	5,908	465	347	9,749	5,684	901	640
54,928	34,527	2,591	2,062	54,841	34,135	2,735	2,163
31,152	19,422	1,854	1,414	31,571	19,392	1,808	1,376
266,419	176,508	19,498	14,890	265,428	176,249	18,828	15,039
45,155	30,257	2,557	2,009	45,723	30,826	2,459	1,903
28,413	20,762	2,705	2,236	28,158	20,677	2,621	2,203
10,273	6,359	573	447	10,461	6,344	851	699
5,849	3,601	910	582	5,857	3,604	1,043	732
160,855	101,953	9,385	6,685	162,667	102,339	9,284	7,051
79,486	49,623	5,627	4,092	81,535	50,205	5,468	4,349
11,924	7,473	807	457	11,721	7,303	918	590
15,199	10,263	733	446	15,295	10,307	990	641
117,290	81,653	7,233	5,734	117,943	81,731	7,053	5,534
60,320	42,801	4,397	3,256	60,467	42,745	4,253	3,490
38,506	28,584	2,512	2,038	38,901	28,472	2,449	1,965
32,238	22,737	1,927	1,394	32,304	22,409	1,865	1,499
41,916	28,175	2,938	2,277	42,179	28,250	2,843	2,210
46,722	27,905	3,269	2,392	47,083	27,324	3,189	2,254
24,055	15,550	836	539	23,718	15,279	1,010	729
45,627	31,886	3,260	2,284	45,429	31,915	3,088	2,247
52,658	37,141	5,432	4,083	52,596	37,577	5,031	3,963
122,195	86,470	5,835	4,182	121,074	84,987	5,836	4,189
77,808	56,694	3,267	2,601	77,323	56,473	3,182	2,556
28,323	16,369	2,393	1,677	28,934	16,612	2,255	1,773
64,541	43,942	4,011	3,193	64,995	43,767	3,890	3,203
14,259	9,271	587	402	14,294	9,087	979	725
24,841	17,694	1,195	1,016	24,677	17,526	1,192	1,008
21,663	13,403	829	692	22,050	13,540	1,101	946
14,974	10,062	849	680	14,913	10,104	1,098	851
71,804	49,594	4,778	3,696	70,886	49,363	4,820	3,711
20,936	12,588	923	575	21,216	12,792	1,031	801
180,144	118,562	13,610	11,079	178,282	117,120	13,017	10,762
83,367	54,072	6,228	4,672	84,535	54,422	6,071	4,722
11,509	8,083	568	474	11,419	7,841	1,060	836
109,120	78,439	7,261	5,705	108,931	77,738	7,248	5,635
46,598	28,847	2,533	2,089	46,622	28,645	2,560	2,085
33,911	23,489	2,017	1,290	34,068	23,687	2,032	1,437

continued

TABLE 2-1 Continued

State	ACS 2005 Housing Units Initial Addresses Selected	ACS 2005 Housing Units Final Interview	ACS 2006 Housing Units Initial Addresses Selected	ACS 2006 Housing Units Final Interview	ACS 2006 Group Quarters Initial Sample Selected	ACS 2006 Group Quarters Final Interview
Pennsylvania	145,000	101,216	143,856	104,132	10,659	7,888
Rhode Island	8,819	6,110	8,720	6,193	1,001	812
South Carolina	41,029	25,642	41,546	26,804	3,313	2,544
South Dakota	11,678	7,969	11,675	8,234	697	589
Tennessee	54,786	36,339	55,342	37,446	3,646	2,903
Texas	203,497	121,858	205,272	129,186	13,872	10,819
Utah	20,545	14,331	20,813	14,909	987	767
Vermont	12,232	7,677	12,143	8,076	541	382
Virginia	61,445	42,957	61,857	44,699	5,647	4,144
Washington	58,811	40,262	58,784	41,301	3,315	2,282
West Virginia	21,128	13,496	20,880	13,871	1,082	793
Wisconsin	82,755	61,063	82,458	62,489	3,786	2,951
Wyoming	6,031	3,877	6,046	3,877	353	247
United States	2,922,656	1,924,527	2,885,384	1,968,362	189,641	145,311

SOURCE: U.S. Census Bureau, data from: http://www.census.gov/acs/www/methodology/sample_size_data/ and http://www.census.gov/acs/www/UseData/sse/.

areas with 65,000 or more people. By 2008, enough responses had been collected to release the 3-year ACS estimates for 2005-2007.

The 3-year estimates cover areas with 20,000 or more people, providing wider tabulation coverage of small geographic areas. By 2010, the first 5-year estimates will have been released, covering 2005-2009. With these estimates, the tabulation coverage of the ACS will have expanded to very small places and neighborhoods, including the areas pertaining to even the smallest local education authorities. Each year, the 1-year, 3-year, and 5-year estimates will be updated to include the most recent data.

In addition to the 1-year, 3-year, and 5-year estimates, the Census Bureau has also released ACS 1-year and 3-year PUMS files, and the 5-year PUMS files are scheduled for release early in 2011. PUMS files contain individual and household records, with confidentiality protected through the following means:

- deleting names and addresses from the records;
- limiting geographic and identification to large areas, known as public-use microdata areas, which are defined to include about 100,000 people; and
- limiting the detail that is provided for sensitive variables: for example, assigning a catchall code to income amounts over a certain threshold, such as $100,000 or more, and not identifying the specific amount.

ACS 2007				ACS 2008			
Housing Units		Group Quarters		Housing Units		Group Quarters	
Initial Addresses Selected	Final Interview	Initial Sample Selected	Final Interview	Initial Addresses Selected	Final Interview	Initial Sample Selected	Final Interview
142,939	102,116	10,572	7,693	141,995	101,559	10,245	7,443
8,654	6,005	965	699	8,636	5,995	990	704
41,878	26,606	3,415	2,708	42,299	26,991	3,312	2,630
11,612	8,000	696	552	11,610	7,853	1,068	866
55,752	37,279	3,590	2,886	56,490	37,688	3,529	2,829
206,891	127,633	13,024	10,556	211,122	127,639	12,522	10,133
21,082	14,854	969	707	21,234	15,060	1,026	736
12,147	7,984	501	409	11,948	7,802	1,030	781
62,090	44,235	5,783	4,197	62,548	44,223	5,731	4,357
58,642	40,886	3,224	2,352	58,805	40,855	3,095	2,260
20,842	13,632	1,132	900	21,028	13,565	1,118	887
81,905	61,524	3,695	2,861	81,123	60,357	3,716	3,050
6,111	3,893	354	262	6,211	3,924	888	672
2,886,453	1,937,659	187,012	142,468	2,894,711	1,931,955	186,862	145,974

ASSESSMENT OF THE DATA

As noted in Chapter 1, Title III of the Elementary and Secondary Education Act requires the U.S. Department of Education (DoEd) to allocate funds to all 50 states, the District of Columbia, and Puerto Rico[2] by a formula in which 80 percent is based on the population of children with limited proficiency in English (relative to national counts of this population). The ACS uses a sample of the population to estimate the number of people with limited English proficiency (LEP).

The definition of the population of children with limited proficiency in English in the ACS derives from the ACS questionnaire which asks the household respondent three questions about the spoken English capability of each household member: see Box 2-1 (also see Chapter 1). The questions are asked of those who are aged 5 years or more. Based on responses to these questions, household members between 5 and 21 years old are categorized as English language learners if the respondent reports that the person speaks a language other than English at home and speaks English less than "very well."

[2]Puerto Rico has a cap; the total amount is not to exceed 0.5 percent of the total amount allotted to all states in a fiscal year.

BOX 2-1
Question on Language Use from the ACS

a. Does this person speak a language other than English at home?
☐ Yes, *go to b*
☐ No, *skip b and c*

b. What is this language?

(For example: Korean, Italian, Spanish, Vietnamese)

c. How well does this person speak English?
☐ Very well
☐ Well
☐ Not well
☐ Not at all

SOURCE: American Community Survey Questionnaire, Form ACS-1 (INFO)(2010)KFI.

Quality of the ACS Language Questions

The ACS questions on English speaking ability evolved directly from similar questions on the former census long form. Indeed, the decennial census has collected information on the ability of the population to speak the English language for well over a century, and the question has evolved over time: see Box 2-2. The census question evolved from a simple English speaking ability question to one which focused on "mother" tongue, and finally in 1980, to the multipart language question that was adopted to fulfill requirements of legislation that sought to identify language limitations which were a source of disadvantage in learning, voting, and access to public services (Kominski, 1989, p. 1). Like other questions on the old census long form, the ones on English speaking ability were incorporated into the ACS during the testing phase and eventually adopted without change. Thus, it is appropriate to review the research used to assess the reasonableness and utility of the language question as it was asked on the decennial census and to compare the estimates of English speaking ability from the census with the estimates from the ACS.

In an article on what "how well" means, Kominski (1989) reported on an independent assessment of English proficiency to validate the multipart question used on the census. Kominski used data from the 1986 National Content Test, a national survey conducted by the Census Bureau to assess new and candidate items for the decennial census. This test included a reinterview survey in which about one-quarter of the original sample was administered follow-up questions. These

BOX 2-2
History of the Census Language Questions

2000: Data collected for all ages, retained for persons 5 years and over
Does this person speak a language other than English at home?
What is this language?
How well does this person speak English (very well, well, not well, not at all)?

1990: Data for persons 5 years and over
Does this person speak a language other than English at home?
What is this language?
How well does this person speak English (very well, well, not well, not at all)?

1980: Asked about persons 3 years and older; tabulated for those 5 years and older
Does this person speak a language other than English at home?
What is this language?
How well does this person speak English (very well, well, not well, not at all)?

1970: No age for question, tabulations limited
What language, other than English, was spoken in this person's home when he was a child?
(Spanish, French, German, Other (specify)_________, None, English only)

1960: Asked of foreign-born persons
What language was spoken in his home before he came to the United States?

1950: No language questions

1940: "Mother tongue (or Native Language)"
Language spoken at home in earliest childhood

1930: Asked of foreign-born persons
"Mother tongue (or Native Language) of Foreign Born"
Language spoken in home before coming to the United States

1920: Place or birth and mother tongue of person and each parent. (May also have used instructions from 1910)

1910: Mother tongue was collected for all foreign-born persons, to be written in with place of birth, also collected for foreign-born parents
Ability to speak English
"Whether able to speak English; or, if not, give language spoken" (specific instructions on correct languages to write and a list of appropriate European languages)

1900: Asked about all persons 10 years and older
"Can you speak English" was asked after two questions on literacy, reading, and writing.

1890: (All persons 10 years and over)
"Able to speak English. If not, the language or dialect spoken." Asked after two questions on literacy, reading, and writing.

1790-1880: No known language or English ability questions

SOURCE: R. Kominski, U.S. Census Bureau (personal communication, June 15, 2009).

questions included items regarding the language spoken in one's home as a child, current number of languages spoken, where they are spoken and with whom, where they were learned, frequency spoken, and four specific skills: ability to read a book in a foreign language, to write a postcard in the language, to read a book in English, and to write a postcard in English. Responses to these questions were used to determine how well the summary evaluations made by a respondent on the single English language proficiency item corresponded to a more detailed assessment of ability.

The study found that women, nonwhite respondents, Hispanics, recent immigrants, and persons with low educational level reported lower English speaking ability than other groups. This group was dominated by recent immigrants, Hispanics, and Spanish speakers. Background factors played a major role in determining the English speaking ability of English among persons who spoke a language other than English. A positive correlation was found between English speaking ability and the language spoken in a person's childhood home. That is, respondents were more likely to indicate that they spoke English "not well" or "not at all" if they resided in homes where a language other than English was spoken. The source from which a person learned a language other than English also appeared to play a role in English speaking ability. When the language was learned at school, military, or somewhere outside of the home, then English speaking ability of the person was almost always "very well." One other finding was that knowing more than one non-English language did not have a significant effect on reported English speaking ability.

The effect of current language use on speaking ability was also investigated. Several findings were reported:

- individuals who spoke a language other than English with friends or at home were more likely to have lower English speaking ability than others;
- individuals who spoke a language other than English either not at all or only at work or school reported speaking English "very well;"
- the frequency with which a language was used influenced the speaking ability of the speaker, so that people who reported speaking ability less than "well" were more likely to use English less frequently or not at all; and
- spoken ability in English was found to be positively correlated with reading and writing ability in English.

However, the data analysis did not provide strong evidence that the language questions are able to differentiate between the lower two levels of English speaking ability ("not well" and "not at all"). In summary, the language question used in the ACS questionnaire was found to have a fair degree of association with a series of other language-related items and to differentiate between the very worst and very best speakers of English language, although it does not distinguish levels of proficiency well among those with limited proficiency.

To assess similarity of the responses to the language item on the census long form to those from the ACS, with its different collection methods, residence rules, and response rates, another study compared national distributions from the ACS, the census, and other Census Bureau surveys for the three items: speaking a language other than English at home, the languages spoken, and English speaking ability (Shin, 2008, p. 2). Results of the 2000 census might not be comparable to the 2005 ACS because of intervening changes in immigration patterns. Hence, Shin compared the census results to those of both the 2005 ACS and the Census 2000 Supplementary Survey (C2SS), a one-time test survey conducted in 1,200 counties that used the same questions as the 2000 census long form, but with operational and data collection methods more similar to those of the ACS.

Box 2-3 compares the item nonresponse rates (as measured by item allocation rates[3]) for the 2000 census, the 2000 C2SS, and the 2005 ACS.

The operational ACS achieved much better response rates for these questions than either the 2000 census or the 2000 C2SS, an important indicator of data quality. The results of the comparison of the 2000 census with the 2000 C2SS are not as clear cut. For example, for those with LEP, the C2SS estimated 19 million people and the 2000 census estimated 21 million people. The census reported a larger percentage of people speaking English less than "very well" than did to the C2SS, but the ACS estimate was much lower than both of them. Shin points out that the differences between the C2SS and the 2000 census may be due to different data collection methods, sample frames, and residence rules, or they may be a product of the intensive campaign that is waged by the decennial census to booster response rates of non-English speaking groups by way of language-based advertising and multiple language questionnaires. For example, the 2000 census was printed in five languages, while the C2SS (and the ACS) were offered only in English and Spanish.

Relationship Between ACS Responses and Tested Proficiency

To evaluate the validity of ACS estimates, it is useful to assess the relationship of responses to the ACS English language proficiency questions with tested proficiency. A study focused on this issue was conducted with data collected in 1982 by the Census Bureau for the DoEd (U.S. Department of Education, 1987). Although dated, the findings of that study are still useful in that they indicate the relationship between the screening questions used in the 1980 census (the first year in which the language ability questions were asked) and an administered English language proficiency test.

The study drew on a special English Language Proficiency Survey, a test that was administered to 8,800 school-age children. A total of 4,000 were from what were then-called "language-minority" households and the remaining 4,800 were from

[3]Item allocation rates are computed as the ratio of the number of eligible persons for which a value was allocated during the editing process for a specific item to the number of people eligible to have responded to that item (Shin, 2008, p. 8).

BOX 2-3
Item Nonresponse Rate Comparisons: 2000 Census, 2000 C2SS, and 2005 ACS (in percentage)

Item	2000 Census	2000 C2SS	2005 ACS
Spoke another language at home	6.2	4.3	1.7
Other language spoken	12.0	8.9	4.0
English speaking ability	8.2	6.0	2.5

NOTE: ACS = American Community Survey, C2SS = Census 2000 Supplementary Survey.
SOURCE: Shin (2008, p. 8).

households in which only English was spoken, as identified in the 1980 census.[4] The test, which was administered by Census Bureau interviewers, was based on a Language Measurement and Assessment Inventory developed earlier for another survey. The test consisted of 10 age-specific tests that were administered at their homes to children aged 5 through 10, with the younger children orally tested with pictures and flash cards, and the older youth taking a written test. Proficiency standards that were applied to the results were the same as the DoEd used for determining the need for special bilingual education services for children in schools.

Test performance was related to a number of variables, including education of the household head, family income, progress in school, nativity and recency of immigration, language spoken in the home, English language ability (using the same indicators as are now used), and membership in specific language groups.

The study compared the numbers of English language proficient children as reported to the DoEd by state educational agencies with estimates derived from the model. The state agency reports identified 1,428,000 children of LEP while the special survey estimated a higher number, 1,752,000. Among the findings of the survey were that about one-third of the school-age children from homes where another language was used some of the time are classified as LEP. The numbers varied widely among the states and across home languages spoken, with households that spoke Spanish and Indo-Asian languages registering the highest test scores for LEP children.

Although, unfortunately, there was not a direct analysis of the state-based relationship between the English speaking ability question and the language proficiency test results, a comparison between the percent of national totals for the estimate

[4]The census households were sampled on the basis of their responses to the questions on spoken language, with language minority households oversampled by a factor of five to develop reliable estimates for the group.

based on speaking English "less than well" and the LEP estimate based on the model showed very consistent patterns for the nine largest LEP population states and for the aggregated group of the rest of the states. However, the total identified by the English speaking ability question on the census was considerably lower than the number identified by the model: 653,600 and 1,752,000, respectively.

It is difficult to state with confidence that the conclusions drawn from this 30-year-old data collection directly bear on the task of this panel. However, it should be observed that, in the only test of the relationship between the English speaking ability question that now appears on the ACS and a somewhat objective test of language proficiency, the patterns of responses between the states seemed to indicate a strong correlation between the results (keeping in mind that the estimate of the number of LEP students was quite different).

Siegel, Martin, and Bruno (2001) reviewed the language questions in the census long form, looking at the conceptual underpinnings of the census data labeled "linguistic isolation."[5] The article presented evidence on non-English language use and analyzed the characteristics of households and areas with high rates of linguistic isolation. In considering the sources of nonsampling error in the language questions of ACS, the authors noted ambiguities of meaning in the question on use of a non-English language at home. Respondents may not know whether to mark "yes" if they practice speaking a language learned in school, speak another language with visitors from outside the country, or engage in other intermittent speech of a non-English language. Alternatively, some may interpret the question to be asking only about habitual speech. "At home" may confuse the respondents if they are recent immigrants who make occasional trips to their home country and speak their native language on those trips. Nonimmigrants can consider "home" to be their childhood home, in which a non-English language might have been spoken. The question may not be applicable to individuals who live alone or do not interact with anybody at home.

Placement of the language item on the questionnaire can influence responses. In 1980, the item nonresponse rate for the language use question was 8.2 percent, and in 1990 it improved to 5.1 percent: the authors attributed the improvement to movement of the item so that it did not immediately follow questions about birthplace and the year a person immigrated to the United States. Such questions evoke an individual's homeland and can cause ambiguity in interpreting the question. The article also pointed out that the subjective character of the English proficiency question makes it vulnerable to a variety of influences, such as situational factors, different reference groups, mode of response (questionnaire or interviewer), and cultural context. Situational factors influence what standards a respondent adopts to judge his or her own English proficiency and that of others in the household.

[5]"Linguistic isolation" is dependent on the English speaking ability of all adults in a household. A household is linguistically isolated if all adults speak a language other than English and none speaks English "very well." Adult is defined as age 14 or older, which identifies household members of high school age and older.

The study found that different reference groups can influence the judgments of proficiency. The standard of comparison might variously be the English speaking ability of native English speakers, of the interviewer, or of members of the respondent's ethnic group or community. The response to the language question varied by survey mode (written or interviewer-administered questionnaire), partly because interviewer administration might induce respondents to provide socially desirable positive answers. If the language question is on the written ACS, individuals can claim any level of proficiency they wish without fear or contradiction. However, during an interview the claim of the individual is put to a test, and such factors as accents of respondents and other irrelevant factors can influence the interviewers' judgments. The respondent's ethnic background may also influence the way proficiency is perceived and reported. McArthur (1991) speculated that Asians systematically underreport English proficiency, while Hispanics overreport it. The study also found the earlier immigrants are less likely to speak a language other than English at home than are more recent immigrants.

Reporting on the ACS is also affected by the person who is selected by the interviewer to respond for the housing unit. By ACS rules, one person may provide data for all members of the household. The household respondent is generally a household member who is at least 18 years old but, if necessary, household members who are 15 and older can be interviewed. Thus, few of the responses concerning English speaking ability are based on student's self-appraisal of their abilities. Typically, the assessment is made by a family member who makes a subjective judgment about the young person's ability.

CONCLUSION 2-1 As a measure to determine Title III allocations, the American Community Survey questions have a number of desirable features, and they provide a uniform estimate across the country. The questions are standardized throughout the country, relatively insensitive to outside influences and transparent in the way that they are used in identifying English language learner students. However, the questions suffer from limitations of conceptual relevance and comprehensiveness of assessment that may affect the validity of estimates based on children's English speaking ability.

ACS ESTIMATES

Numbers

The ACS estimates in this section are derived from special tabulations prepared for the use of the panel by the Census Bureau under the sponsorship of the DoEd.

Table 2-2 shows ACS 1-year estimates of ELL children and youth along with the corresponding standard errors and coefficients of variation[6] for 50 states and District of Columbia. The second column reports the 1-year estimates for 2005. For example, in 2005, there were 18,745 ELL students aged 5-21 in the state of Alabama. The subsequent columns report the ELL estimates for the years 2006, 2007, and 2008.

Table 2-3 reports the ACS 3-year estimates of ELL children and youth. The column "2005-2007" represents the average number of ELL children and youth for the 3-year period. Table 2-3 also presents the respective standard errors and coefficients of variation of the 3-year estimates to provide the viewers a more comprehensive view of the reliability of the estimates.

Shares and Ratios

As noted above, Title III funds are allocated to each state (and the District of Columbia) on the basis of their shares of the national total: those percentage shares are shown in Table 2-4. To show the variability of ACS share estimates, Table 2-5 presents the standard errors for the share allocations.

The count and share of ELL students in a state are proportional to the product of the total school population in the state and the percentage of ELL students. The latter percentage characterizes the concentration of ELL students in the state. This within-state percentage can be said to reflect a state's burden; that is, the percentage of its school-age population that needs Title III services. This measure is useful in making comparisons among states that are independent of the size of the state.

For purposes of this report, we define the percentage of children and youth who receive Title III services as the ratio of ELL children and youth to all children and youth in each state. More specifically, the ratio is that of the ACS estimate of the population of ELL children and youth aged 5-18 years old enrolled in public school and the ACS estimate of all such children and youth aged 5-18: see Table 2-6. For example, the ACS estimate for 2005 indicates that 1.15 percent of the school-age children in public schools in Alabama were ELL students.

PROPERTIES OF THE ESTIMATES

Precision, Reliability, and Stability

Because the ACS surveys a sample of the population, estimates based on it are subject to random variation (sampling error). The amount of random variation in the estimates can itself be estimated, and is commonly summarized by standard error (a measure of how large sampling error would typically be for a given estimate) or coefficient of variation (the standard error of the estimate divided by the magnitude

[6]The coefficient of variation is computed as the ratio of the standard deviation to the mean. It compares the degree of variation from one data series to another, and allows for the influence of different means.

TABLE 2-2 English Language Learning Children and Youth Aged 5-21, by State, 2005-2008

State	ACS 2005			ACS 2006			ACS 2007			ACS 2008		
	EST	SE	CV	EST	SE	CV	EST	SE	CV	EST	SE	CV
Alabama	18,745	1,806	0.10	20,740	1,639	0.08	21,725	1,928	0.09	18,055	1,443	0.08
Alaska	4,225	902	0.21	6,400	799	0.12	7,015	1,031	0.15	4,740	559	0.12
Arizona	121,895	4,702	0.04	135,310	4,718	0.03	141,980	5,701	0.04	131,480	4,796	0.04
Arkansas	17,095	1,432	0.08	17,565	1,433	0.08	18,280	1,661	0.09	17,230	1,499	0.09
California	1,097,205	16,272	0.01	1,038,305	11,935	0.01	1,003,915	11,379	0.01	948,515	12,331	0.01
Colorado	61,675	3,116	0.05	65,380	3,532	0.05	67,120	3,373	0.05	59,030	3,076	0.05
Connecticut	33,165	2,383	0.07	32,420	2,262	0.07	25,870	1,803	0.07	24,770	1,754	0.07
Delaware	8,355	802	0.10	7,340	915	0.12	6,900	1,017	0.15	5,625	875	0.16
District of Columbia	3,490	617	0.18	3,955	735	0.19	3,385	735	0.22	2,700	619	0.23
Florida	234,505	7,672	0.03	236,570	6,917	0.03	231,810	5,808	0.03	213,005	6,449	0.03
Georgia	85,275	3,514	0.04	91,010	3,587	0.04	92,605	3,434	0.04	76,245	3,186	0.04
Hawaii	14,230	1,660	0.12	12,900	1,406	0.11	10,745	1,102	0.10	16,865	1,919	0.11
Idaho	9,860	1,215	0.12	10,880	1,283	0.12	10,340	1,127	0.11	11,285	1,222	0.11
Illinois	182,730	6,211	0.03	175,625	5,652	0.03	178,480	5,381	0.03	169,395	4,835	0.03
Indiana	40,740	2,204	0.05	41,135	2,236	0.05	37,395	2,143	0.06	39,705	1,942	0.05
Iowa	16,015	1,081	0.07	18,510	1,410	0.08	15,415	1,235	0.08	15,440	1,325	0.09
Kansas	21,115	1,455	0.07	20,405	1,683	0.08	19,820	1,310	0.07	20,165	1,845	0.09
Kentucky	17,160	1,515	0.09	16,625	1,244	0.07	20,830	1,786	0.09	18,255	1,588	0.09
Louisiana	14,165	1,353	0.10	13,440	1,304	0.10	15,425	1,321	0.09	17,445	1,364	0.08
Maine	3,535	693	0.20	4,620	833	0.18	3,755	726	0.19	2,650	466	0.18
Maryland	47,550	2,819	0.06	42,010	2,213	0.05	46,010	2,350	0.05	40,730	2,549	0.06
Massachusetts	64,815	4,140	0.06	67,250	2,791	0.04	61,345	2,884	0.05	63,520	2,766	0.04
Michigan	62,675	2,904	0.05	57,345	2,629	0.05	57,275	2,451	0.04	52,615	2,869	0.05
Minnesota	39,575	2,251	0.06	45,730	2,783	0.06	42,200	2,056	0.05	46,910	2,629	0.06
Mississippi	7,870	1,175	0.15	7,725	915	0.12	8,100	780	0.10	8,035	902	0.11
Missouri	21,765	2,003	0.09	24,400	2,025	0.08	28,095	2,024	0.07	24,775	1,818	0.07

Montana	2,185	522	0.24	2,010	472	0.23	2,240	419	0.19	2,280	484	0.21
Nebraska	14,935	1,242	0.08	16,930	1,365	0.08	14,080	1,335	0.09	14,305	1,386	0.10
Nevada	38,540	2,669	0.07	43,680	2,437	0.06	46,440	2,416	0.05	49,670	2,526	0.05
New Hampshire	5,000	806	0.16	3,200	594	0.19	4,050	694	0.17	3,925	644	0.16
New Jersey	107,955	3,620	0.03	104,210	3,394	0.03	97,980	3,877	0.04	101,215	3,697	0.04
New Mexico	28,805	2,298	0.08	34,825	3,041	0.09	27,700	1,770	0.06	24,925	2,353	0.09
New York	275,230	7,116	0.03	302,040	6,232	0.02	279,875	6,728	0.02	290,170	7,273	0.03
North Carolina	70,970	4,095	0.06	85,770	3,482	0.04	79,025	2,892	0.04	83,400	3,584	0.04
North Dakota	1,700	388	0.23	2,210	553	0.25	2,660	523	0.20	2,440	499	0.20
Ohio	48,005	2,530	0.05	47,905	2,328	0.05	44,645	2,756	0.06	47,275	2,963	0.06
Oklahoma	21,085	1,781	0.08	20,205	1,293	0.06	20,595	1,667	0.08	18,995	1,379	0.07
Oregon	49,910	3,066	0.06	45,650	2,724	0.06	47,150	2,612	0.06	41,520	2,616	0.06
Pennsylvania	74,245	3,602	0.05	68,215	3,426	0.05	70,835	2,923	0.04	71,820	3,237	0.05
Puerto Rico	835,520	5,343	0.01	845,825	4,945	0.01	841,715	5,308	0.01	820,655	4,956	0.01
Rhode Island	12,130	1,687	0.14	9,260	980	0.11	10,510	1,340	0.13	10,880	1,147	0.11
South Carolina	22,940	1,518	0.07	24,430	1,771	0.07	23,810	1,914	0.08	22,000	2,005	0.09
South Dakota	4,065	993	0.24	3,255	492	0.15	2,620	545	0.21	2,805	741	0.26
Tennessee	28,635	2,156	0.08	28,460	1,968	0.07	31,520	2,025	0.06	28,925	2,230	0.08
Texas	570,145	9,866	0.02	586,090	8,899	0.02	599,265	9,096	0.02	595,070	10,881	0.02
Utah	21,050	1,626	0.08	28,115	2,036	0.07	29,035	1,954	0.07	27,080	2,220	0.08
Vermont	1,900	430	0.23	1,515	401	0.26	1,565	355	0.23	1,725	345	0.20
Virginia	57,440	2,645	0.05	65,565	3,296	0.05	49,795	2,537	0.05	54,860	2,783	0.05
Washington	78,270	3,068	0.04	80,355	3,707	0.05	87,725	4,165	0.05	85,105	3,164	0.04
West Virginia	3,250	526	0.16	3,935	647	0.16	3,565	494	0.14	3,275	638	0.19
Wisconsin	38,855	1,957	0.05	39,655	2,100	0.05	43,430	2,022	0.05	35,845	1,912	0.05
Wyoming	2,130	516	0.24	1,625	372	0.23	1,875	414	0.22	1,475	401	0.27
United States	3,828,820	25,849	0.01	3,862,675	20,298	0.01	3,797,820	20,240	0.01	3,670,185	23,813	0.01

NOTES: CV = coefficients of variation; EST = estimated number; SE = standard error.

TABLE 2-3 Average Number of ELL Children and Youth Aged 5-21, by State

	ACS 2005-2007			ACS 2006-2008		
State	Estimate	SE	CV	Estimate	SE	CV
Alabama	19,295	865	0.04	18,665	766	0.04
Alaska	5,915	496	0.08	6,170	425	0.07
Arizona	132,520	2,906	0.02	134,520	2,549	0.02
Arkansas	18,185	869	0.05	17,360	797	0.05
California	1,045,820	6,993	0.01	988,085	6,728	0.01
Colorado	63,905	1,643	0.03	63,210	1,969	0.03
Connecticut	31,060	1,207	0.04	28,020	1,066	0.04
Delaware	7,530	526	0.07	6,565	520	0.08
District of Columbia	3,785	375	0.10	2,950	376	0.13
Florida	233,140	3,732	0.02	224,250	3,081	0.01
Georgia	89,105	1,986	0.02	84,940	1,973	0.02
Hawaii	12,465	815	0.07	13,160	832	0.06
Idaho	11,215	789	0.07	11,180	702	0.06
Illinois	179,805	3,433	0.02	172,420	2,855	0.02
Indiana	39,085	1,170	0.03	38,755	1,225	0.03
Iowa	16,910	798	0.05	16,745	724	0.04
Kansas	20,780	1,084	0.05	19,690	1,038	0.05
Kentucky	19,225	839	0.04	18,885	890	0.05
Louisiana	15,760	886	0.06	16,375	930	0.06
Maine	4,125	418	0.10	3,870	488	0.13
Maryland	45,820	1,489	0.03	43,625	1,317	0.03
Massachusetts	65,915	1,906	0.03	63,735	1,856	0.03
Michigan	60,600	1,797	0.03	55,390	1,496	0.03
Minnesota	43,365	1,534	0.04	45,155	1,381	0.03
Mississippi	8,805	606	0.07	8,755	581	0.07
Missouri	25,695	1,160	0.05	25,985	1,100	0.04
Montana	2,295	287	0.13	2,495	265	0.11
Nebraska	15,150	699	0.05	14,870	719	0.05
Nevada	43,395	1,600	0.04	46,525	1,464	0.03
New Hampshire	4,695	513	0.11	3,845	348	0.09
New Jersey	103,225	1,887	0.02	100,645	2,315	0.02
New Mexico	29,900	1,366	0.05	28,455	1,358	0.05
New York	289,480	3,977	0.01	290,395	4,000	0.01
North Carolina	76,535	1,778	0.02	79,945	1,899	0.02
North Dakota	2,165	286	0.13	2,190	269	0.12
Ohio	47,580	1,425	0.03	46,095	1,344	0.03
Oklahoma	21,325	752	0.04	20,140	1,010	0.05
Oregon	47,585	1,480	0.03	44,605	1,484	0.03
Pennsylvania	71,770	1,672	0.02	70,115	1,868	0.03
Rhode Island	10,725	809	0.08	10,195	680	0.07
South Carolina	24,255	1,051	0.04	23,715	1,127	0.05
South Dakota	3,480	348	0.10	3,165	399	0.13
Tennessee	30,675	1,252	0.04	29,770	940	0.03
Texas	581,800	6,085	0.01	586,510	5,692	0.01
Utah	26,535	1,298	0.05	27,745	1,304	0.05
Vermont	1,755	213	0.12	1,510	187	0.12
Virginia	57,335	1,754	0.03	56,330	1,467	0.03
Washington	80,445	1,953	0.02	82,905	2,178	0.03
West Virginia	4,120	423	0.10	3,870	416	0.11
Wisconsin	41,555	1,168	0.03	39,205	1,223	0.03
Wyoming	1,980	248	0.13	1,825	192	0.11
United States	3,839,580	13,565	0.004	3,745,540	15,296	0.004

TABLE 2-4 Percentage Share of ELL Children and Youth Aged 5-21, by State

	ACS 2005	ACS 2006	ACS 2007	ACS 2008	ACS 2005-2007	ACS 2006-2008
State	Share	Share	Share	Share	Share	Share
Alabama	0.49	0.54	0.57	0.49	0.50	0.50
Alaska	0.11	0.17	0.18	0.13	0.15	0.16
Arizona	3.18	3.50	3.74	3.58	3.45	3.59
Arkansas	0.45	0.45	0.48	0.47	0.47	0.46
California	28.66	26.88	26.43	25.84	27.24	26.38
Colorado	1.61	1.69	1.77	1.61	1.66	1.69
Connecticut	0.87	0.84	0.68	0.67	0.81	0.75
Delaware	0.22	0.19	0.18	0.15	0.20	0.18
District of Columbia	0.09	0.10	0.09	0.07	0.10	0.08
Florida	6.12	6.12	6.10	5.80	6.07	5.99
Georgia	2.23	2.36	2.44	2.08	2.32	2.27
Hawaii	0.37	0.33	0.28	0.46	0.32	0.35
Idaho	0.26	0.28	0.27	0.31	0.29	0.30
Illinois	4.77	4.55	4.70	4.62	4.68	4.60
Indiana	1.06	1.06	0.98	1.08	1.02	1.03
Iowa	0.42	0.48	0.41	0.42	0.44	0.45
Kansas	0.55	0.53	0.52	0.55	0.54	0.53
Kentucky	0.45	0.43	0.55	0.50	0.50	0.50
Louisiana	0.37	0.35	0.41	0.48	0.41	0.44
Maine	0.09	0.12	0.10	0.07	0.11	0.10
Maryland	1.24	1.09	1.21	1.11	1.19	1.16
Massachusetts	1.69	1.74	1.62	1.73	1.72	1.70
Michigan	1.64	1.48	1.51	1.43	1.58	1.48
Minnesota	1.03	1.18	1.11	1.28	1.13	1.21
Mississippi	0.21	0.20	0.21	0.22	0.23	0.23
Missouri	0.57	0.63	0.74	0.68	0.67	0.69
Montana	0.06	0.05	0.06	0.06	0.06	0.07
Nebraska	0.39	0.44	0.37	0.39	0.39	0.40
Nevada	1.01	1.13	1.22	1.35	1.13	1.24
New Hampshire	0.13	0.08	0.11	0.11	0.12	0.10
New Jersey	2.82	2.70	2.58	2.76	2.69	2.69
New Mexico	0.75	0.90	0.73	0.68	0.78	0.76
New York	7.19	7.82	7.37	7.91	7.54	7.75
North Carolina	1.85	2.22	2.08	2.27	1.99	2.13
North Dakota	0.04	0.06	0.07	0.07	0.06	0.06
Ohio	1.25	1.24	1.18	1.29	1.24	1.23
Oklahoma	0.55	0.52	0.54	0.52	0.56	0.54
Oregon	1.30	1.18	1.24	1.13	1.24	1.19
Pennsylvania	1.94	1.77	1.87	1.96	1.87	1.87
Rhode Island	0.32	0.24	0.28	0.30	0.28	0.27
South Carolina	0.60	0.63	0.63	0.60	0.63	0.63
South Dakota	0.11	0.08	0.07	0.08	0.09	0.08
Tennessee	0.75	0.74	0.83	0.79	0.80	0.79
Texas	14.89	15.17	15.78	16.21	15.15	15.66
Utah	0.55	0.73	0.76	0.74	0.69	0.74
Vermont	0.05	0.04	0.04	0.05	0.05	0.04
Virginia	1.50	1.70	1.31	1.49	1.49	1.50
Washington	2.04	2.08	2.31	2.32	2.10	2.21
West Virginia	0.08	0.10	0.09	0.09	0.11	0.10
Wisconsin	1.01	1.03	1.14	0.98	1.08	1.05
Wyoming	0.06	0.04	0.05	0.04	0.05	0.05

SOURCE: U.S. Census Bureau Special Tabulations.

TABLE 2-5 Standard Errors of Percentage Shares of ELL Children and Youth Aged 5-21, by State (in percentage)

	ACS 2005	ACS 2006	ACS 2007	ACS 2008	ACS 2005-2007	AC 2006-2008
State	SE of Share	SE of Share	SE of Share	SE of Share	SE of Share	SE of Share
Alabama	0.05	0.04	0.05	0.04	0.02	0.02
Alaska	0.02	0.02	0.03	0.02	0.01	0.01
Arizona	0.12	0.12	0.15	0.13	0.07	0.07
Arkansas	0.04	0.04	0.04	0.04	0.02	0.02
California	0.38	0.27	0.26	0.29	0.15	0.14
Colorado	0.08	0.09	0.09	0.08	0.04	0.05
Connecticut	0.06	0.06	0.05	0.05	0.03	0.03
Delaware	0.02	0.02	0.03	0.02	0.01	0.01
District of Columbia	0.02	0.02	0.02	0.02	0.01	0.01
Florida	0.20	0.18	0.15	0.17	0.09	0.08
Georgia	0.09	0.09	0.09	0.09	0.05	0.05
Hawaii	0.04	0.04	0.03	0.05	0.02	0.02
Idaho	0.03	0.03	0.03	0.03	0.02	0.02
Illinois	0.16	0.14	0.14	0.13	0.09	0.07
Indiana	0.06	0.06	0.06	0.05	0.03	0.03
Iowa	0.03	0.04	0.03	0.04	0.02	0.02
Kansas	0.04	0.04	0.03	0.05	0.03	0.03
Kentucky	0.04	0.03	0.05	0.04	0.02	0.02
Louisiana	0.04	0.03	0.03	0.04	0.02	0.02
Maine	0.02	0.02	0.02	0.01	0.01	0.01
Maryland	0.07	0.06	0.06	0.07	0.04	0.03
Massachusetts	0.11	0.07	0.08	0.07	0.05	0.05
Michigan	0.08	0.07	0.06	0.08	0.05	0.04
Minnesota	0.06	0.07	0.05	0.07	0.04	0.04
Mississippi	0.03	0.02	0.02	0.02	0.02	0.02
Missouri	0.05	0.05	0.05	0.05	0.03	0.03
Montana	0.01	0.01	0.01	0.01	0.01	0.01
Nebraska	0.03	0.04	0.04	0.04	0.02	0.02
Nevada	0.07	0.06	0.06	0.07	0.04	0.04
New Hampshire	0.02	0.02	0.02	0.02	0.01	0.01
New Jersey	0.09	0.09	0.10	0.10	0.05	0.06
New Mexico	0.06	0.08	0.05	0.06	0.04	0.04
New York	0.18	0.16	0.17	0.19	0.10	0.10
North Carolina	0.11	0.09	0.08	0.10	0.05	0.05
North Dakota	0.01	0.01	0.01	0.01	0.01	0.01
Ohio	0.07	0.06	0.07	0.08	0.04	0.04
Oklahoma	0.05	0.03	0.04	0.04	0.02	0.03
Oregon	0.08	0.07	0.07	0.07	0.04	0.04
Pennsylvania	0.09	0.09	0.08	0.09	0.04	0.05
Rhode Island	0.04	0.03	0.04	0.03	0.02	0.02
South Carolina	0.04	0.05	0.05	0.05	0.03	0.03
South Dakota	0.03	0.01	0.01	0.02	0.01	0.01
Tennessee	0.06	0.05	0.05	0.06	0.03	0.02
Texas	0.24	0.22	0.22	0.28	0.15	0.14
Utah	0.04	0.05	0.05	0.06	0.03	0.03
Vermont	0.01	0.01	0.01	0.01	0.01	0.00
Virginia	0.07	0.08	0.07	0.08	0.05	0.04
Washington	0.08	0.10	0.11	0.08	0.05	0.06
West Virginia	0.01	0.02	0.01	0.02	0.01	0.01
Wisconsin	0.05	0.05	0.05	0.05	0.03	0.03
Wyoming	0.01	0.01	0.01	0.01	0.01	0.01

SOURCE: U.S. Census Bureau Special Tabulations.

TABLE 2-6 Ratio of ELL Students Aged 5-18 in Public Schools to All Students Aged 5-18 in Public Schools (in percentage)

	ACS 2005	ACS 2006	ACS 2007	ACS 2008	ACS 2005-2007	ACS 2006-2008
State	Ratio	Ratio	Ratio	Ratio	Ratio	Ratio
Alabama	1.15	1.44	1.36	1.31	1.29	1.33
Alaska	2.40	3.34	3.56	3.09	3.24	3.53
Arizona	7.89	8.50	8.82	8.01	8.43	8.40
Arkansas	2.29	2.40	2.54	1.94	2.51	2.32
California	12.00	11.34	11.13	10.54	11.50	11.00
Colorado	4.97	5.55	5.53	5.14	5.32	5.37
Connecticut	3.48	3.90	2.91	2.55	3.44	3.16
Delaware	3.45	3.51	3.20	2.93	3.35	3.26
District of Columbia	3.63	2.91	2.60	2.57	2.93	2.34
Florida	5.59	5.35	5.33	4.99	5.42	5.16
Georgia	3.13	3.24	3.31	2.89	3.24	3.15
Hawaii	5.05	4.40	3.40	5.60	4.21	4.36
Idaho	2.47	2.69	2.30	2.84	2.71	2.70
Illinois	5.91	5.41	5.72	5.57	5.67	5.54
Indiana	2.13	2.24	1.98	2.12	2.10	2.13
Iowa	2.45	2.24	2.07	2.03	2.25	2.22
Kansas	3.10	2.75	2.60	2.85	2.88	2.81
Kentucky	1.42	1.18	1.48	1.67	1.44	1.54
Louisiana	1.39	0.96	1.44	1.40	1.42	1.29
Maine	1.16	1.81	1.35	1.05	1.47	1.55
Maryland	3.41	2.85	3.25	2.87	3.19	3.02
Massachusetts	4.42	4.51	4.16	4.15	4.36	4.27
Michigan	2.54	2.27	2.23	2.06	2.39	2.18
Minnesota	3.29	3.52	3.67	4.07	3.54	3.81
Mississippi	0.70	0.87	0.79	0.74	0.92	0.98
Missouri	1.35	1.57	1.88	1.63	1.68	1.75
Montana	0.98	0.94	0.86	0.66	1.06	1.02
Nebraska	3.31	3.56	3.22	3.40	3.36	3.42
Nevada	5.73	5.99	6.45	7.62	6.26	6.79
New Hampshire	1.70	0.70	1.35	1.09	1.46	1.09
New Jersey	4.77	4.68	4.26	4.65	4.60	4.56
New Mexico	5.92	7.03	6.26	5.43	6.34	6.14
New York	5.82	5.72	5.59	5.87	5.68	5.72
North Carolina	3.09	3.54	3.19	3.73	3.30	3.47
North Dakota	1.24	0.95	1.70	1.44	1.30	1.27
Ohio	1.45	1.43	1.29	1.42	1.40	1.38
Oklahoma	2.33	1.84	2.17	1.89	2.10	2.02
Oregon	6.30	5.47	5.40	4.90	5.69	5.28
Pennsylvania	2.34	2.05	2.11	2.22	2.17	2.13
Rhode Island	4.43	3.11	4.71	4.52	3.99	4.12
South Carolina	2.07	1.84	2.03	1.77	2.05	1.99
South Dakota	2.20	1.06	0.99	1.66	1.40	1.38
Tennessee	1.81	1.62	1.76	1.71	1.75	1.76
Texas	9.73	9.69	10.02	10.07	9.76	9.89
Utah	2.89	3.54	3.52	3.15	3.37	3.42
Vermont	1.17	1.03	0.77	0.85	0.95	0.87
Virginia	2.92	3.28	2.51	2.94	2.91	2.92
Washington	5.43	4.94	5.64	5.60	5.26	5.38
West Virginia	0.72	1.13	0.87	0.79	0.99	0.99
Wisconsin	2.93	2.82	3.39	2.45	3.11	2.91
Wyoming	1.44	1.17	0.82	1.34	1.17	1.20

SOURCE: U.S. Census Bureau Special Tabulations.

of the estimate). Smaller standard errors indicate greater precision of estimation. The standard errors for estimates of the state percentages of the total ELL population (corresponding to the percentages in Table 2-4) are shown in Table 2-5, above. The standard errors for each state's estimated ratios (percentages of the school-age population that are ELL students, in Table 2-6) are shown in Table 2-7.

The precision of estimates for small areas (those with low population sizes) or small population groups can be problematic in any survey. The ACS estimation program addresses this problem by combining data across consecutive years, producing 3- and 5-year estimates as well as those based on a single year of data. Because these estimates are based on more data, they are more precise than 1-year estimates, with smaller standard errors. Because the first 5-year estimates (which will have even smaller standard errors than the 3-year estimates) were not scheduled for release until late 2010, they are not considered in this report.

The 1-year estimates are released for every state, but the sample for some of the smaller states may be so small that the sampling error is substantial, especially for relatively small population subgroups, such as the ELL population. But the 3-year estimates, though more precise, will be slower to respond to changes in the size of the ELL population in each state. The tradeoff of these conflicting values is considered in the following comparison of the accuracy of the 1- and 3-year estimates.

In Table 2-5 (above), the mean of standard errors for all 1-year estimates of the state percentage of the national ELL count is 0.07 percent; the corresponding mean for 3-year estimates is 0.04 percent. Table 2-7 shows the mean of standard errors for all 1-year estimates is 0.28 percent; and the corresponding mean for 3-year estimates is 0.16 percent. Thus, by combining data for 3 years, the standard error is cut almost in half.

To give a sense of the effects of random errors on allocations to states, consider the coefficients of variation (CV) of estimates of state numbers of ELL students, shown in Table 2-2 (above). Table 2-8 shows the statistics for all states and for groups of states classified by their overall share of ELLs as large, medium, small, and "minimum." (Variations in the shares of ELL children and youth of the "minimum" states do not affect their allocations, as they generally fall below the $500,000 threshold.)

The overall mean CV is 0.091 (9.1%) for 1-year estimates and .051 (5.1%) for 3-year estimates. As expected, the mean CV is smaller in the larger states because the samples in these states are generally larger. There are very large coefficients of variation for the "minimum" states, but these are irrelevant for policy since these states almost always receive the minimum allocation of $500,000 (or slightly more). At the other extreme, coefficients of variation for the large states, accounting for about 73 percent of all ELL children, are quite small, around 3 percent for 1-year estimates and 1.7 percent for 3-year estimates. The most problematic group is the small states, with fairly large coefficients of variation (around 22% for 1-year estimates and 12% for 3-year estimates). That is, there is substantial relative variation in estimated allocation for these states, but they account for only about 4 percent of all ELL children, spread over 15 states, so the allocation amounts at stake are very small.

TABLE 2-7 Standard Errors of Ratio of ELL Students Aged 5-18 in Public School to All Students Aged 5-18 in Public School (in percentage)

	ACS 2005	ACS 2006	ACS 2007	ACS 2008	ACS 2005-2007	ACS 2006-2008
State	SE of Ratio	SE of Ratio	SE of Ratio	SE of Ratio	SE of Ratio	SE of Ratio
Alabama	0.12	0.18	0.18	0.13	0.09	0.08
Alaska	0.67	0.45	0.55	0.38	0.34	0.30
Arizona	0.36	0.38	0.42	0.35	0.22	0.18
Arkansas	0.21	0.23	0.28	0.23	0.15	0.13
California	0.19	0.16	0.14	0.16	0.09	0.09
Colorado	0.33	0.33	0.30	0.34	0.18	0.20
Connecticut	0.30	0.31	0.25	0.22	0.15	0.15
Delaware	0.46	0.64	0.62	0.59	0.34	0.33
District of Columbia	0.82	0.84	0.83	0.84	0.40	0.44
Florida	0.25	0.20	0.19	0.16	0.13	0.09
Georgia	0.16	0.17	0.15	0.17	0.10	0.11
Hawaii	0.76	0.50	0.46	0.69	0.36	0.31
Idaho	0.32	0.38	0.31	0.36	0.26	0.21
Illinois	0.25	0.21	0.20	0.20	0.13	0.12
Indiana	0.15	0.14	0.15	0.13	0.08	0.09
Iowa	0.21	0.18	0.21	0.20	0.12	0.12
Kansas	0.28	0.25	0.22	0.29	0.18	0.18
Kentucky	0.15	0.13	0.16	0.16	0.10	0.09
Louisiana	0.19	0.15	0.16	0.14	0.10	0.09
Maine	0.31	0.40	0.31	0.21	0.19	0.20
Maryland	0.23	0.20	0.21	0.22	0.12	0.13
Massachusetts	0.29	0.28	0.24	0.25	0.15	0.15
Michigan	0.14	0.12	0.12	0.12	0.09	0.07
Minnesota	0.21	0.25	0.24	0.26	0.15	0.15
Mississippi	0.15	0.12	0.13	0.12	0.08	0.09
Missouri	0.17	0.16	0.17	0.14	0.11	0.10
Montana	0.24	0.27	0.20	0.22	0.16	0.15
Nebraska	0.34	0.40	0.40	0.39	0.16	0.20
Nevada	0.42	0.44	0.40	0.51	0.28	0.26
New Hampshire	0.31	0.21	0.28	0.22	0.21	0.13
New Jersey	0.21	0.19	0.20	0.23	0.11	0.13
New Mexico	0.57	0.74	0.48	0.60	0.34	0.35
New York	0.18	0.15	0.17	0.19	0.09	0.10
North Carolina	0.23	0.18	0.16	0.19	0.11	0.12
North Dakota	0.29	0.26	0.43	0.39	0.21	0.24
Ohio	0.10	0.09	0.10	0.11	0.06	0.06
Oklahoma	0.26	0.17	0.22	0.19	0.10	0.11
Oregon	0.47	0.39	0.34	0.37	0.22	0.18
Pennsylvania	0.16	0.14	0.12	0.14	0.07	0.08
Rhode Island	0.74	0.44	0.81	0.64	0.35	0.36
South Carolina	0.19	0.19	0.21	0.21	0.11	0.13
South Dakota	0.67	0.25	0.28	0.52	0.20	0.22
Tennessee	0.18	0.15	0.17	0.18	0.10	0.08
Texas	0.18	0.16	0.17	0.20	0.13	0.10
Utah	0.26	0.31	0.29	0.33	0.19	0.21
Vermont	0.28	0.39	0.24	0.21	0.16	0.15
Virginia	0.18	0.21	0.18	0.17	0.12	0.10
Washington	0.26	0.29	0.33	0.24	0.16	0.17
West Virginia	0.14	0.22	0.18	0.16	0.14	0.13
Wisconsin	0.20	0.19	0.21	0.17	0.11	0.11
Wyoming	0.40	0.37	0.28	0.42	0.19	0.19

SOURCE: U.S. Census Bureau Special Tabulations.

TABLE 2-8 Coefficients of Variation of Estimates of ELL Students, by State Size

State Share	ACS 2005	ACS 2006	ACS 2007	ACS 2008	ACS 2005-2007	ACS 2006-2008
Large	0.030	0.028	0.029	0.029	0.017	0.016
Medium	0.063	0.060	0.057	0.063	0.035	0.035
Small	0.125	0.115	0.117	0.117	0.069	0.067
Minimum	0.226	0.219	0.209	0.230	0.117	0.119
All	0.095	0.089	0.088	0.093	0.0513	0.0510

NOTES:
Large States: Arizona, California, Florida, Georgia, Illinois, New Jersey, New York, Texas, Washington.
Medium States: Colorado, Connecticut, Indiana, Kansas, Maryland, Massachusetts, Michigan, Minnesota, Missouri, North Carolina, New Mexico, Nevada, Pennsylvania, Oregon, Ohio, Oklahoma, South Carolina, Tennessee, Utah, Virginia, Wisconsin.
Small States: Alabama, Alaska, Arkansas, Delaware, Hawaii, Idaho, Iowa, Kentucky, Louisiana, Maine, Mississippi, Nebraska, New Hampshire, Rhode Island, West Virginia.
Minimum States: District of Columbia, Montana, North Dakota, South Dakota, Vermont, Wyoming.

Another way to assess reliability of the estimates is by the intergeographic unit reliability of estimates of the percentage of school-enrolled children who are classified ELL students (the ELL rate), which places large and small states on a comparable scale. This statistic summarizes on a scale from 0 to 1 how well the data distinguish states by this measure of burden: 0 means the data are completely unreliable, equivalent to random noise, and 1 means that the data have no error and all differences among state estimates are due only to actual differences among their populations. Technically, reliability for state *s* is given by the formula

$$\text{reliability} = \sigma^2/(V_s + \sigma^2),$$

where σ^2 is between-state model variance (estimated using a hierarchical model[7]) and V_s is the sampling variance of the estimate for state *s*. In each of the years from 2006 to 2008, reliabilities for 1-year ACS estimates range from 0.88 in the least reliably measured states (generally, small states) to over 0.99 for the most reliably measured states. These statistics indicate that the ACS is precise enough to distinguish well among states with low and high rates of ELL students.

We next considered the reliability of estimates of *changes* in this ratio, using the same hierarchical estimation model and formula but applying it to differences between consecutive years (from 2006 to 2007 and from 2007 to 2008). In either pair of years, the model estimates that interyear changes in rates (after removing the

[7]"The hierarchical model is of the form $y_i = \beta_0 + \beta_1 x_i + u_i + e_i$, where y_i is the ACS estimate of a rate for state *i*, and x_i is the corresponding rate from state-provided data. Random effects $u_i \sim N(0,\sigma^2)$ and $e_i \sim N(0,V_i)$ are respectively model and sampling error for the ACS estimates in state *i*, and V_i is the sampling variance of y_i."

average national trend) were quite small, with standard deviations of approximately 0.14 and 0.15 percent. The 1-year estimates were not sufficiently precise to reliably assess these generally small changes, with reliabilities in the two intervals ranging from below 0.02 for the least precisely measured states to a high of about 0.55 for the most precisely measured ones. (The latter number indicates that about half of the variation in estimates of change for states with large samples is due to random sampling variation rather than actual year-to-year change.) This finding strongly suggests that it is futile to attempt to use the 1-year ACS estimates to track annual changes, except perhaps when a state has an exceptionally large change in its ELL student population.

Finally, the 1- and 3-year estimates were compared with regard to the *stability* over time of the estimated shares (see Table 2-9).

When summarized by the sum of the absolute differences in the ratios of ELL children and youth in the various states, the sum of changes is much larger (6.23%, 5.03%, and 5.26%, respectively, for 2005-2006, 2006-2007, and 2007-2008) for the 1-year ACS estimates than for the 3-year estimates (3.07% for 2005-2007 to 2006-2008). This result is as anticipated because of the overlap of consecutive 3-year estimates. For example, considering the difference of the 2005-2007 and 2006-2008 estimates, two-thirds of the data (2006 and 2007 data) are identical in the two estimates so the difference is only one-third of the difference between the 2005 and 2008 estimates. Thus, use of 3-year ACS estimates automatically makes estimates more stable, though at the cost of slower responsiveness to robust changes in the size of the ELL population in any state, because any sharp change in the ELL population would only be reflected in one-third of the next year's estimate. Given the importance of stability of funding share over periods of a few years, the DoEd would be well advised to use the 3-year ACS estimates rather than the 1-year estimates, and to consider use of the 5-year ACS estimates when they become available and their statistical properties are investigated. (We present more detailed information on stability, with comparisons to state-based estimates, in Chapter 5.)

CONCLUSION 2-2 Allocations based on 3-year American Community Survey (ACS) estimates are substantially more precise and stable, especially in states with relatively small populations, than those based on 1-year ACS estimates. Neither 1-year nor 3-year ACS estimates can precisely estimate annual changes in English language learner rates, but use of 3-year estimates smooths variation over time.

Sensitivity to Variation in Subpopulations

It is useful to examine the extent to which ACS estimates of states' shares change when the criteria used to define ELL status are modified, perhaps because of limitations in some of the data sources. Currently, the ELL group is defined as 5- to 21-years-olds who speak English less than very well. Because the ACS is a popula-

TABLE 2-9 Absolute Difference in Percentage Share of States Across Years (in percentage)

State	ACS 2006 Compared with ACS 2005 Absolute Difference	ACS 2007 Compared with ACS 2006 Absolute Difference	ACS 2008 Compared with ACS 2007 Absolute Difference	ACS 2006-2008 Compared with ACS 2005-2007 Absolute Difference
Alabama	0.047	0.035	0.080	0.004
Alaska	0.055	0.019	0.056	0.011
Arizona	0.319	0.235	0.156	0.140
Arkansas	0.008	0.027	0.012	0.010
California	1.776	0.446	0.590	0.858
Colorado	0.082	0.075	0.159	0.023
Connecticut	0.027	0.158	0.006	0.061
Delaware	0.028	0.008	0.028	0.021
District of Columbia	0.011	0.013	0.016	0.020
Florida	0.000	0.021	0.300	0.085
Georgia	0.129	0.082	0.361	0.053
Hawaii	0.038	0.051	0.177	0.027
Idaho	0.024	0.009	0.035	0.006
Illinois	0.226	0.153	0.084	0.080
Indiana	0.001	0.080	0.097	0.017
Iowa	0.061	0.073	0.015	0.007
Kansas	0.023	0.006	0.028	0.016
Kentucky	0.018	0.118	0.051	0.003
Louisiana	0.022	0.058	0.069	0.027
Maine	0.027	0.021	0.027	0.004
Maryland	0.154	0.124	0.102	0.029
Massachusetts	0.048	0.126	0.115	0.015
Michigan	0.152	0.024	0.075	0.099
Minnesota	0.150	0.073	0.167	0.076
Mississippi	0.006	0.013	0.006	0.004
Missouri	0.063	0.108	0.065	0.025
Montana	0.005	0.007	0.003	0.007
Nebraska	0.048	0.068	0.019	0.002
Nevada	0.124	0.092	0.131	0.112
New Hampshire	0.048	0.024	0.000	0.020
New Jersey	0.122	0.118	0.178	0.001
New Mexico	0.149	0.172	0.050	0.019
New York	0.631	0.450	0.537	0.214
North Carolina	0.367	0.140	0.192	0.141
North Dakota	0.013	0.013	0.004	0.002
Ohio	0.014	0.065	0.113	0.009
Oklahoma	0.028	0.019	0.025	0.018
Oregon	0.122	0.060	0.110	0.048
Pennsylvania	0.173	0.099	0.092	0.003

TABLE 2-9 Continued

State	ACS 2006 Compared with ACS 2005 Absolute Difference	ACS 2007 Compared with ACS 2006 Absolute Difference	ACS 2008 Compared with ACS 2007 Absolute Difference	ACS 2006-2008 Compared with ACS 2005-2007 Absolute Difference
Rhode Island	0.077	0.037	0.020	0.007
South Carolina	0.033	0.006	0.028	0.001
South Dakota	0.022	0.015	0.007	0.006
Tennessee	0.011	0.093	0.042	0.004
Texas	0.282	0.606	0.434	0.506
Utah	0.178	0.037	0.027	0.050
Vermont	0.010	0.002	0.006	0.005
Virginia	0.197	0.386	0.184	0.011
Washington	0.036	0.230	0.009	0.118
West Virginia	0.017	0.008	0.005	0.004
Wisconsin	0.012	0.117	0.167	0.036
Wyoming	0.014	0.007	0.009	0.003
United States	6.230	5.027	5.264	3.066

SOURCE: U.S. Census Bureau Special Tabulations.

tion survey, one can examine the sensitivity of the allocations when the criteria are altered slightly. For example, if the goal is to align the ACS data more closely to state counts of ELL children and youth, the group of interest would be those aged 5-18 and enrolled in public school. We examined the effects of changing the criteria in terms of age (5-18 versus 5-21), enrollment status (all enrolled students versus those in public schools only), and English speaking ability (speak English less than very well versus speak English less than well).

This analysis was conducted using the 3-year ACS estimates for 2006-2008, with the following steps:

1. We selected as the base definition those aged 5-21 and speaking English less than very well. We calculated the state shares using this definition.
2. We then varied the definition and calculated the revised state shares.
3. We then calculated and summarized differences. Suppose A_x is the state share for state X under the base criteria, and B_x the state's share under revised criteria (e.g., when the age range is restricted to 5-18). The difference of the two shares is $(B_x - A_x)$. We then took the absolute value of the difference to obtain the *absolute difference* and summarized these values by their mean across states, as the the *mean absolute difference* (MAD).

4. We also calculated and summarized *relative difference*. This was calculated by dividing the absolute difference by the average of the two shares, $(B_x - A_x)/((B_x + A_x)/2)$. We then took the mean of these values to calculate the *mean absolute relative difference* (MARD). The MAD tends to be heavily influenced by differences in *large* states, the MARD gives comparatively more weight to smaller states.

The results of these analyses are presented in Table 2-10. The first row shows the effect of changing the age range to 5-18. The second row shows the effect of restricting students enrolled in any kind of school. The third row shows the results when school enrollment is restricted to those in public school.[8] The fourth row shows the results when both criteria are applied—restricting the population to 5- to 18-year-olds enrolled in public school.

In this summary table, we report the statistics for all states and for groups of states classified by their overall share of ELLs under the base allocation as large, medium, small, and "minimum."[9] As noted above, variations in the shares of ELL children and youth of the "minimum" states do not affect their allocations, as they generally fall below the $500,000 threshold.

As can be seen in Table 2-11, the variations in age criteria did not influence the allocation of states very much (MAD, 0.06%; MARD, 1.04%). The allocations are more sensitive to restricting estimates to children and youth enrolled in schools (MAD, 0.07%; MARD, 5.46%), and even more so to restricting to those enrolled in public schools (MAD, 0.14%; MARD, 7.92%). Thus, with the latter restriction, states would on the average see a noticeable change (7.92%) in their allocations. This presumably reflects some differences in school enrollment rates among ELL children and youth. The combined restriction by both the age and public school enrollment criteria has a slightly larger effect on allocations (MAD, 0.16%; MARD, 9.58%).

For each of the revisions of criteria we considered, the MAD, reflecting the amount of money that would be moved, is largest for the large states (those with the biggest shares of the national population of ELL children and youth). However, the relative impact (measured by the MARD), reflecting the percentage by which a revision would modify a state's allocation, tends to be larger for the medium and small states, for which a small amount of money can be a large percentage of a state's allocation. The biggest relative changes are in the "minimum" states, but these would not affect their allocations because they receive a fixed amount.

[8]The comparison is only for public schools because state estimates are only available for students in public schools.

[9]The large states are Arizona, California, Florida, Georgia, Illinois, New Jersey, New York, Texas, and Washington. The medium states are Colorado, Connecticut, Indiana, Kansas, Maryland, Massachusetts, Michigan, Minnesota, Missouri, North Carolina, New Mexico, Nevada, Pennsylvania, Oregon, Ohio, Oklahoma, South Carolina, Tennessee, Utah, Virginia, and Wisconsin. The small states are Alabama, Alaska, Arkansas, Delaware, Hawaii, Idaho, Iowa, Kentucky, Louisiana, Maine, Mississippi, Nebraska, New Hampshire, Rhode Island, and West Virginia. Minimum allocation states are the District of Columbia, Montana, North Dakota, South Dakota, Vermont, and Wyoming.

TABLE 2-10 Difference in Percentage Share of ELL Students of States by Varying Age Groups, Enrollment Status, and Type of School (in percentage)

Base Category: Children and Youth Aged 5-21 Who Speak English Less Than "Very Well"

Alternatives to Base Category	Mean Absolute Difference in Share[a]	Mean Absolute Relative Difference[b]
Age Group:		
5-18 years old		
All	**0.06**	**1.04**
Large	0.21	0.75
Medium	0.03	0.74
Small	0.01	1.24
Minimum	0.00	2.01
Enrollment Status:		
Enrolled in School		
All	**0.07**	**5.46**
Large	0.26	4.23
Medium	0.06	5.08
Small	0.02	6.61
Minimum	0.00	5.77
Type of School:		
Public Schools		
All	**0.14**	**7.92**
Large	0.57	6.57
Medium	0.07	6.61
Small	0.02	9.11
Minimum	0.01	11.52
5-18 Years Old,		
Public Schools:		
All	**0.16**	**9.58**
Large	0.67	7.45
Medium	0.08	7.49
Small	0.03	10.91
Minimum	0.01	16.74

[a]The mean absolute difference in share is calculated by taking an average of absolute difference in share of all states and group of states.

[b]The mean absolute relative difference in share is calculated by taking an average of absolute relative difference in share of all states and group of states.

CONCLUSION 2-3 The 3-year American Community Survey (ACS) estimates of English language learner (ELL) children and youth are relatively insensitive to definitional changes in age range, but they are sensitive to changes in enrollment status and type of school. Consequently, adjusting the age group used in the ACS definition of ELL children and youth from 5-21 years of age to 5-18 years of age will have little effect on the percentage share of Title III funds going to the states, but changing the

TABLE 2-11 Difference in Percentage Share of ELL Students of States by Varying ELL Criterion

Alternatives to Base Category	Mean Absolute Difference in Share	Mean Absolute Relative Difference	Overall Rate
Base Category: Children and Youth Aged 5-21 Who Speak English Less Than Very Well			
Speaking English Less Than Well			
All	**0.17**	**11.71**	**38.58**
Large	0.59	5.37	37.82
Medium	0.11	9.71	40.14
Small	0.05	16.82	42.16
Minimum	0.01	15.40	40.57
Base Category: Children and Youth Aged 5-18 Public School Enrolled Who Speak English Less Than Very Well			
Speaking English Less Than Well			
All	**0.27**	**16.00**	**30.88**
Large	1.07	7.82	29.77
Medium	0.14	13.21	33.54
Small	0.06	21.44	35.13
Minimum	0.02	24.43	40.27

NOTES:
Large States: Arizona, California, Florida, Georgia, Illinois, New Jersey, New York, Texas, Washington.
Medium States: Colorado, Connecticut, Indiana, Kansas, Maryland, Massachusetts, Michigan, Minnesota, Missouri, North Carolina, New Mexico, Nevada, Pennsylvania, Oregon, Ohio, Oklahoma, South Carolina, Tennessee, Utah, Virginia, Wisconsin.
Small States: Alabama, Alaska, Arkansas, Delaware, Hawaii, Idaho, Iowa, Kentucky, Louisiana, Maine, Mississippi, Nebraska, New Hampshire, Rhode Island, West Virginia.
Minimum States: District of Columbia, Montana, North Dakota, South Dakota, Vermont, Wyoming.

enrollment status definition to limit the group to public school children and youth would have a measurable effect on the shares. In this regard, the ACS measure is more closely aligned with the statutory language than are the figures provided by state education authorities.

Sensitivity to Variations in Cut Points

Another sensitivity analysis considered the allocation effects of alternative ACS proficiency cut points. Currently, an English language learner is defined as one who speaks English "less than very well." Using special tabulations provided by the Census Bureau, we examined the impact of changing the proficiency criterion to "less than well," which has the effect of considering those who speak English "well" as proficient rather than nonproficient. This effect was examined under two different assumptions about the age range and school enrollment criteria corresponding to the base category (5- to 21-year-olds) and last rows (5- to 18-year-olds in public schools) of Table 2-9.

The results, presented in Table 2-10, show that ACS estimates are more sensitive to this change of cut point (MAD, 0.17%; MARD, 11.71%) than to changes in age range and enrollment status. The impact is even greater with the stricter age and enrollment criteria (MAD, 0.27%; MARD. 16.00%). This result is not surprising given that those speaking English "less than well" constitute only about one-third of those speaking English "less than very well" (39% in the less restrictive age enrollment criteria; 31% with the more restrictive criteria). Given the variation in ethnic composition, country of origin, and recency of immigration of the immigrant populations of the various states, the distribution of ELL children and youth across the nonproficient categories on the ACS is likely to vary as well.

In view of the strong sensitivity of the estimates of ELL students to the cut points selected, the continued use of "less than very well" as the cutoff used in ACS to define English language learners is warranted. This determination is consistent with evidence cited earlier in this chapter that even though the language question in ACS is not able to precisely distinguish between the four categories of English speakers, it does differentiate between the worst and best speakers of English language.

CONCLUSION 2-4 The American Community Survey estimates of English language learner (ELL) children and youth are very sensitive to cutoff points in the ELL definition. Changing the criterion from "less than very well" to "less than well" can bring about substantial changes in a state's share of the total number of ELL children and youth, and, consequently, in the state's allocation.

We return to this topic in Chapter 5, which presents further evidence bearing on the choice of cut point.

Reporting of Type of School

The ACS asks whether each student attends "public" or "private" school. We know of no assessment of the accuracy of the responses to this question. In particular, charter schools are regarded as public schools for statistical purposes, but because they are often regarded by parents as an alternative to regular district-administered schools, they might be misreported as private. This reporting could affect estimates of public school ELL rates if charter schools have different rates of ELL enrollment than district-administered schools, but it would affect neither estimates of total ELL students nor those of total ELL children.

Coverage Error

The ACS provides yearly survey data on important economic and social characteristics of the U.S. population, but the definition of that population has changed over time in ways that have introduced coverage error. The ACS for 2005 covered the household population, while the 2006, 2007, and 2008 ACS covered not only the household population, but also people who live in college dormitories, armed forces barracks, prisons, nursing homes, correctional institutions, and other group quarters.[10] The decision to include or exclude housing units of a certain type introduces coverage error. There are two kinds of coverage error: undercoverage (when housing units or people do not have a chance of being selected in the sample) and overcoverage (when housing units or people have more than one chance of being selected in the sample or are included in the sample when they should not have been).[11] If the characteristics of undercovered or overcovered housing units or individuals differ from those that are selected, the ACS may not provide an accurate picture of the population.

ACS reduces coverage error by controlling specific survey estimates to independent population controls[12] by sex, age, race, and Hispanic origin for population estimates and to independent housing unit controls for housing unit estimates. The Census Bureau calculates coverage rates to measure coverage error in the ACS, and these rates are weighted to reflect the probability of selection into the sample, the subsampling for personal visit follow-up, and nonresponse. As the coverage rate drops below 100 percent, the weights of the people in the survey need greater adjustment in the final weighting procedure to reach the independent estimate. If the rate is greater than 100 percent, the ACS population estimates are downweighted to match the independent estimates. Independent population estimates are produced by the Census Bureau using independent data on such characteristics as housing, births, deaths, and immigration. The base for these independent estimates is the decennial census.

The coverage rates for housing units, group quarters, and the total population for 2005-2008 are shown in Table 2-12. The coverage rate for the total population for 2008 was 93.8 percent, and that for the Hispanic population was 92.5 percent. On the basis of these data, it can be postulated that coverage error is not a significant concern for the ELL estimates.

---

[10]Residences that are not in ACS but were part of the census long-form sample are circus quarters, crews on merchant ships, domestic violence shelters, recreational vehicles in campground, soup kitchen or mobile food van sites, and street location for the homeless.

[11]Overcoverage occurs when units or people have multiple chances of selection; for example, addresses listed more than once on the frame, or people included on a household roster at two different sampled addresses. For details see: Census Bureau, ACS Design and Methodology, Chapter 15, http://www.census.gov/acs/www/Downloads/survey_methodology/acs_design_methodology_ch15.pdf [December 2010].

[12]The use of population controls can introduce another source of error (National Research Council, 2007, pp. 201-208).

TABLE 2-12 Coverage Rates for Housing Units, Group Quarters, and Total Population (in percentage*)

	Housing Units	Groups Quarters Population	Total Population								
Year	Total	Total	Total	Male	Female	White Non-Hispanic	Black Non-Hispanic	American Indian and Alaska Native Non-Hispanic	Asian Non-Hispanic	Native Hawaiian and Other Pacific Islander Non-Hispanic	Hispanic
2008	98.7	80.8	93.8	92.6	95.0	94.7	89.7	96.2	96.9	85.8	92.5
2007	98.5	79.6	94.2	93.2	95.2	95.4	89.1	96.8	95.6	96.1	92.8
2006	98.7	76.2	94.4	93.4	95.3	95.6	89.6	98.0	93.4	93.0	92.9
2005	98.5	N/A	95.1	93.9	96.2	96.3	90.7	97.9	94.5	84.0	93.6

*The Census Bureau does not calculate coverage rates of gender groups cross-tabulated by racial groups (e.g., white non-Hispanic male).

SOURCE: Data from http://www.census.gov/acs/www/acs-php/quality_measures_coverage_2008.php [June 2010].

TABLE 2-13 Allocation Rates for Language Questions in ACS, for United States* (in percentage)

Item	2008	2007	2006	2005
Speaks another language at home total population 5 years and over	3.1	2.2	2.0	1.7
Language spoken total population 5 years and over who speak another language at home	5.3	4.4	4.2	4.0
English ability total population 5 years and over who speak another language at home	3.9	3.1	2.8	2.5

*The item allocation rates for year 2005 are for housing units only. The item allocation rates for 2006 to 2008 include housing units and group quarters populations.
SOURCE: Census Bureau Quality Measures Page, available: http://www.census.gov/acs/www/UseData/sse/ita/ita_def.htm [accessed May 2010].

Nonresponse Error

The population of interest under Title III is a relatively small subgroup of the population, and the quality of the data for this group is very sensitive to item nonresponse to the questions that are used as criteria for the ELL definition. The Census Bureau does adjust for nonresponse, using methods of imputation that fall into two categories: "assignment," using the a response to one question that implies the value for a missing response to another question, and "allocation," using statistical procedures such as within-household or nearest-neighbor matrices populated by donors. Item nonresponse is measured through the calculation of an allocation rate. The formula for allocation rate[13] of an item (A) for a particular state (x) in a year (y) is given as follows:

$$\text{Allocation rate for item A (state x, year y)} = \left(\frac{\text{total number of responses allocated for item A in state x for year y)}}{\text{total number of responses to item A required in state x for year y}} \right) * 100$$

The allocation rate for United States is calculated by summing over the total number of responses allocated and responses required for an item across all states. The overall item allocation rate for the questions determining ELL status for 2005-2008 is from the Census Bureau.[14]

As shown in Table 2-13, the number of responses allocated or imputed re-

[13]From the Census Bureau, see http://www.census.gov/acs/www/UseData/sse/ita/ita_def.htm [May 2010].

[14]The item allocation rates for 2005 are for housing units only; the item allocation rates for 2006 to 2008 include housing units and group quarters populations.

sponses for "Speaks another language at home," "Language spoken," and "English ability" items are very low.

We note that the amount of imputation over the period from 2005 to 2007 for all items has increased, which relates to the issue of response rate to surveys in general. The amount of imputation is also of concern because it introduces a variability that is not currently factored into the estimates of sampling errors from the ACS (National Research Council, 2007, p. 254).

CONCLUSION 2-5 Item nonresponse is a troublesome and growing issue for items used in the calculation of the number of English language learner children and youth.

3

Quality and Comparability of State Tests of English Language Proficiency

The No Child Left Behind (NCLB) Act of 2001 requires states to annually assess the English language proficiency of their students who are classified as limited English proficiency (LEP), also referred to as English language learner (ELL) students. The law (Title III) requires states to establish English language proficiency (ELP) content standards and to use a single ELP test to assess students' progress in and mastery of these standards in four domains: reading, writing, speaking, and listening. Results from the annual administration of ELP tests are used to report on students' progress in and attainment of English language proficiency. The tests may also be used to identify ELL students and to determine when they should end ELL status, often in conjunction with other criteria.

In this chapter, we discuss the ELP tests that states use and compare and contrast their features. We examine the technical quality of the tests, not with the intent of doing a full-scale evaluation of each of them, but rather to consider their use in classifying ELL students and measuring students' progress in learning English. We reviewed the tests by examining the information reported in their technical manuals and supplementary reports with regard to how the tests were developed, the skills that they measure, how the test scores are derived and reported, the reliability of those scores, and the validity of the decisions based on the scores. We consider these aspects of the tests in relation to established technical standards for developing tests, such as those published in the *Standards for Educational and Psychological Testing* (American Educational Research Association, American Psychological Association, and National Council on Measurement in Education, 1999). We focus primarily on the extent to which the tests are likely to support valid decisions about students' English proficiency and the comparability of those decisions across states, given the available data. More detailed information on the tests is in Appendix A.

NCLB REQUIREMENTS FOR ENGLISH LANGUAGE PROFICIENCY TESTS

ELP tests have long been used by the states to classify ELL students by language proficiency level for instructional program placement and decision-making purposes. Many were developed in response to legislation and litigation of the 1970s (e.g., the *Lau v. Nichols* Supreme Court decision and the Equal Educational Opportunities Act of 1974), a time when very few instruments were available to assess ELP (Bauman et al., 2007). For the most part, these tests reflected the predominant structural linguistic approach to assessing ELP (Abedi, 2007; Francis and Rivera, 2007). They were designed to assist local educators with English as a second language and bilingual education program placement and exit decisions, and they typically focused on oral (listening and speaking) domains, measuring discrete phonological and basic interpersonal communication skills. These tests focused largely on basic interpersonal communication skills rather than academic language skills. As a result, students may have scored well on them without having mastered the English language skills needed for learning subject matter in an English-only classroom (Lara et al., 2007).

Before NCLB, there was no attempt to bring uniformity to the ELP assessments with regard to what they measured, their technical measurement properties, or how they were used. Moreover, states typically allowed local school districts to choose among a variety of commercial ELP assessments that varied widely in their characteristics, emphases, and technical properties. Reviews of the pre-NCLB ELP tests have revealed that they differed from each other in their theoretical foundations, the type of language assessed, the types of skills assessed (i.e., receptive or expressive skills), the content assessed, the types of assessment tasks, structural characteristics (i.e., administration procedures, grade level ranges, assessment time required), and technical qualities (e.g., reliability and validity) (Del Vecchio and Guerrero, 1995; Zehler et al., 1994). Many of these tests were not based on an operationally defined concept of ELP, had limited questions that measured academic language proficiency, were not based on explicitly articulated ELP content standards, and had psychometric flaws and other shortcomings (Abedi, 2007, 2008; Bauman et al., 2007; Del Vecchio and Guerrero, 1995; Lara et al., 2007; Zehler et al., 1994).

Under Titles I and III of NCLB, the U.S. Department of Education (DoEd) required states to make improvements to ELP assessments, specifically (adapted from Abedi, 2008, p. 5):

(1) Develop and implement ELP standards suitable for ELL students learning English as a second language.
(2) Implement a single reliable and valid ELP assessment that is aligned to ELP standards and that annually measures listening, speaking, reading, writing, and comprehension skills.
(3) Align the ELP test with the state's challenging academic content and student academic achievement standards described in section 1111(b)(1)(PL 107-110. Available: http://www2.ed.gov/policy/elsec/leg/esea02/index.html [April 2011]).

(4) Establish two annual measurable achievement objectives for ELL students that explicitly define, measure, and report on the student's expected progress toward and attainment of ELP.

These requirements brought about significant changes in the states' ELP tests, and therefore the tests currently used differ in a number of important ways from the pre-NCLB tests (Abedi, 2007; Bauman et al., 2007; Francis and Rivera, 2007; Lara et al., 2007; Rebarbar et al., 2007). First, the new ELP tests are standards based. This means that the first step in the assessment development process is to identify and adopt a set of ELP content standards. Then test specifications are developed to guide test item development in each of the four major language domains (reading, writing, listening, and speaking). Test items are then designed to measure a representative sample of the standards. Although the new ELP tests are not tests of academic content, they are intended to assess the types of language skills required for students to access the core academic content.[1] In line with NCLB, the new ELP tests measures both receptive (listening, reading) and expressive (speaking, writing) language proficiency skills and comprehension.[2] They also more explicitly link and assess skills related to English as a second language and academic language skills, required to be successful in school (for details on the academic language construct, see Anstrom et al., 2009; Bailey and Heritage, 2008, or Scarcella, 2008).

The new ELP assessments offer different forms of the test for each cluster of grades (e.g., early elementary, later elementary, middle school, high school), which are designed to measure growth in ways that reflect the increasing complexity of given language proficiency levels at different age/grade levels. For example, what constitutes intermediate-level academic oral language skill for a 3rd-grade student may be quite different from that for an 8th-grade student. Pre-NCLB assessments generally clustered large numbers of grade levels together. A last major difference is that, unlike the pre-NCLB tests, the new tests are designed for high-stakes decision making and are treated as secure assessments. These changes in the tests have been judged to represent a significant departure from prior practices (Bauman et al., 2007; Lara et al., 2007; Mathews, 2007; Rebarbar et al., 2007).

STATE ENGLISH LANGUAGE PROFICIENCY TESTS

Development

To develop the tests required by NCLB, the DoEd provided grants under Title VI (Section 6112) of the act. The grants allowed for development, validation, and implementation of ELP assessments and encouraged states to work together in

[1]In other words, the assessment should evaluate the language skill (i.e., vocabulary, structure, grammar) needed to access the content of the core academic content standards.

[2]We note that there were proficiency assessments in the 1980s that measured skills in these domains, but they were not standards based.

consortia. In a second round of funding, the DoEd provided additional support for some of the consortia to field test and validate the assessments. Under the grant competition, four different consortia of states were formed, and most of the states initially joined one of these groups.

One consortium was led by the Council of Chief State School Officers with states in the Limited English Proficient State Collaborative on Assessment and Student Standards (LEP-SCASS), which developed the English Language Development Assessment (ELDA).[3] Initially, 18 states were members of the LEP-SCASS, and 14 states participated in the process of developing, field testing, validating, and implementing ELDA as an operational assessment (Saez, program director, Council of Chief State School Officers, personal communication, August 4, 2010).[4]

Another consortium funded by the DoEd initially included three states (Alabama, Delaware, and Wisconsin) and was led by the Wisconsin Department of Public Instruction.[5] Shortly after being funded, seven additional states joined the consortium (Alabama, District of Columbia, Illinois, Maine, New Hampshire, Rhode Island, and Vermont). Now known as the World-Class Instructional Design and Assessment (WIDA) Consortium, this effort produced the assessment called Assessing Comprehension and Communication in English State to State for English Language Learners (or more simply, the ACCESS). Both the LEP-SCASS and the WIDA consortia continue to work actively with state constituents in administering and refining the assessments. As of March 2010 the WIDA Consortium included 23 states, and the LEP-SCASS included 7 states. Membership in these two consortia is dynamic, with new states joining the consortia on an on-going basis.

Two other state consortia initially funded by the DoEd are no longer active, although they made considerable progress in developing test items. The Mountain West Assessment Consortium (MWAC) included 11 states, led by the Utah State Office of Education.[6] The MWAC's assessment was not fully operational when the grant expired, and the consortium's test item bank was subsequently made available to the member states. Three states (Idaho, Montana, and Utah) used the item bank and incorporated the consortium's test questions into their state proficiency assessments.

---

[3]This consortia worked in collaboration with the American Institutes for Research and with Measurement Incorporated, with external advice from the Center for the Study of Assessment Validity and Evaluation at the University of Maryland (see Lara et al., 2007).

[4]Nevada led the collaboration, with Georgia, Indiana, Iowa, Kentucky, Louisiana, Nebraska, Nevada, New Jersey, Ohio, Oklahoma, South Carolina, Virginia, and West Virginia. See Lara et al. (2007) for a more complete history of this consortium's development efforts.

[5]The Wisconsin department worked in collaboration with the Center for Applied Linguistics, the University of Wisconsin system, and the University of Illinois. See Bauman et al. (2007) for a more complete history of this consortium's development efforts.

[6]The other states were Alaska, Colorado, Idaho, Michigan, Montana, Nevada, New Mexico, North Dakota, Oregon, Utah, and Wyoming. The Utah State Office of Education collaborated with Measured Progress as the test developer. See Mathews (2007) for a more complete history of this consortium's development efforts.

The other consortium originally funded by the DoEd was English Proficiency for All Students, which included five states and was led by Accountability Works.[7] This consortium produced the Comprehensive English Language Learning Assessment (CELLA), which is now used only by Florida.

In addition to state consortia funded through the DoEd, commercial test publishers also developed ELP assessments that met the requirements of NCLB. For instance, CTB/McGraw Hill, which had previously developed an assessment called the Language Assessment Scales (LAS), created the Language Assessment Scales Links K-12 (LAS Links), and concordance tables were produced so that scores on the LAS could be converted to the score scale used for LAS Links. Harcourt, Inc. (now Pearson) developed the new Stanford English Language Proficiency Test (SELP). Some states decided to use one of these commercially developed tests. Typically, the test publisher worked with the state to customize ("augment") the assessment so that it was better aligned with the state's ELP content standards and met the state's needs. In a similar vein, some states created customized versions of consortia-developed tests. For instance, Ohio created its own test (the OTELA) derived from the test item bank and scales of the ELDA (American Institutes for Research).[8] Other states—including some states with the largest ELL enrollments—developed their own unique ELP test: examples include the California English-Language Development Test, the New York English as a Second Language Achievement Test, Oregon's web-based English Language Proficiency Assessment, and the Texas English Language Proficiency Assessment System.

English Language Proficiency Assessments Used by the States

NCLB initially required states to establish ELP standards and implement an ELP assessment aligned to these standards by the 2002-2003 school year (U.S. Department of Education, 2010b, p. 8).[9] This presented a considerable challenge to states, and many participated in one of the four consortia as they worked to develop their standards or assessments and meet the federal deadlines (U.S. Department of Education, 2010b, p. 9). In the end, some adopted the consortium-based assessment, some adopted the consortium's standards, and some adapted consortium standards for their own needs.[10]

---

[7]The consortium partners were Florida, Maryland, Michigan, Pennsylvania, and Tennessee, working in collaboration with the Educational Testing Service. See Rebarber et al. (2007) for a more complete history of this consortium's development efforts.

[8]OTELA is actually a shortened version of ELDA. It was developed to reduce the administration time required for ELDA and to reduce the emphasis on entry level skills while maintaining acceptable levels of reliability and validity.

[9]On July 1, 2005, the deadline was extended to the spring of the 2005-2006 school year (U.S. Department of Education, 2005a, p. 23 of Title III Policy: State of the States); http://wvconnections.k12.wv.us/documents/Timelinefor ELPAssessment.doc.)

[10]For instance, in 2004-2005, 38 state Title III directors indicated that they were participating in one of the four consortia to develop standards or assessments (U.S. Department of Education, 2010b, p. 9).

Table 3-1 shows the test used by each state for the 2009-2010 school year, as reported by Title III officials in each state: 23 of the states use ACCESS (an increase from 15 states in 2005-2006); 7 use ELDA; and 4 use LAS Links, with augmentation as needed to address the state's standards. Two states use augmented versions of the SELP assessment, published by Pearson. The remaining 15 states use a unique test (including California, New York, Oregon, Texas) or a test derived from a consortium test (e.g., Ohio). Thus, in the 2009-2010 school year, the states used approximately 19 different proficiency assessments. However, a simple count of the number of different tests (based on their names) overstates their differences because of the specificity in the federal requirements and the extent of collaboration among states and consortia, as well as private developers, to meet those requirements.

Tests Selected for Panel Review

An in-depth review of all of the state tests was beyond the scope of time and resources available for our study. We therefore identified a subset of the tests to review. For efficiency, we first identified the tests used by more than one state, which include the ACCESS, the ELDA, LAS Links, and SELP. We wanted to be sure to include the tests used in states with large numbers of ELL students, so we next rank ordered the states according to the numbers of ELL students, identifying the 10 states that reported the highest numbers of ELL students over the past 5 years (in order by volume):[11] California, Texas, Florida, New York, Illinois, Arizona, North Carolina, Colorado, Virginia, and Washington. Together, these 10 states account for approximately 75 percent of the ELL students in the country—roughly 3.4 million students.

California, Florida, New York, and Texas—the states with the highest numbers of ELL students—each use their own state-developed tests. The other six states are either members of the WIDA Consortium that uses the ACCESS test, or they use an augmented version of the SELP or LAS Links. Thus, in this chapter, we review the overall technical characteristics and comparability of eight tests, four used by multiple states (ACCESS, ELDA, LAS Links, and SELP) and four used by a single state (CELDT used in California, TELPAS used in Texas, CELLA used in Florida, and NYSESLAT used in New York). Because many other states use one of the tests in the first group, our review covers the tests used by 40 states. Table 3-2 lists the tests that we reviewed and shows the states that use them.

In reviewing these assessments, we gathered general information about each test (e.g., number of subtests, types of questions, scores derived, and proficiency standards). We examined information reported in their technical manuals and supplementary reports with regard to the ways that the tests were developed, the skills that they measure, the ways that the scores are derived and reported, the reliability of those scores, and the validity of the decisions based on the scores. In conducting the review, we examined the materials for evidence that the information was pro-

[11] See Chapters 4 and 5 for further details about numbers of ELL students per state.

TABLE 3-1 English Language Proficiency Assessments, by State, 2009-2010 School Year

State	English Language Proficiency Assessment
Alabama	Accessing Comprehension and Communication in English State to State (ACCESS)
Alaska	IDEA Proficiency Test (IPT)
Arizona	Arizona English Language Learner Assessment (AZELLA) (customized version of the SELP)
Arkansas	English Language Development Assessment (ELDA)
California	California English Language Development Test (CELDT)
Colorado	Colorado English Language Assessment (CELA) (customized version of LAS Links)
Connecticut	Language Assessment Scales Links (LAS Links)
Delaware	ACCESS
District of Columbia	ACCESS
Florida	Comprehensive English Language Learning Assessment (CELLA)
Georgia	ACCESS
Hawaii	ACCESS
Idaho	Idaho English Language Assessment (IELA) (items drawn from MWAC item bank)
Illinois	ACCESS
Indiana	LAS Links
Iowa	ELDA
Kansas	Kansas English Language Proficiency Assessment (KELPA)
Kentucky	ACCESS
Louisiana	ELDA
Maine	ACCESS
Maryland	LAS Links
Massachusetts	Massachusetts English Proficiency Assessment-Reading and Writing (MEPA-R/W) and Massachusetts English Language Assessment-Oral (MELA-O)
Michigan	Michigan English Language Proficiency Assessment (MI-ELPA) (items initially drawn from MWAC and SELP item banks)
Minnesota	K-2 Reading and Writing Checklist Test of Emerging Academic English (TEAE) (grades 3-12) Minnesota Modified Student Oral Language Observation Matrix (MN-SOLOM) (grades K-12)
Mississippi	ACCESS
Missouri	ACCESS
Montana	MontCAS English Language Proficiency Assessment (MontCAS ELP) (adapted items from MWAC)
Nebraska	ELDA
Nevada	Nevada State English Language Proficiency Assessment (NV-ELPA)
New Hampshire	ACCESS
New Jersey	ACCESS
New Mexico	ACCESS
New York	New York State English as a Second Language Achievement Test (NYSESLAT) (items initially drawn from SELP item bank)

continued

TABLE 3-1 Continued

State	English Language Proficiency Assessment
N. Carolina	ACCESS
N. Dakota	ACCESS
Ohio	Ohio Test of Language Acquisition (OTELA) (modified version of ELDA)
Oklahoma	ACCESS
Oregon	Oregon English Language Proficiency Assessment (OR-ELPA)
Pennsylvania	ACCESS
Rhode Island	ACCESS
S. Carolina	ELDA
S. Dakota	ACCESS
Tennessee	Tennessee English Language Placement Assessment (TELPA)
Texas	Texas English Language Proficiency Assessment Systems (TELPAS)
Utah	Utah Academic Language Proficiency Assessment (UALPA) (adapted items from MWAC)
Vermont	ACCESS
Virginia	ACCESS
Washington	Washington Language Proficiency Test II (WLPT–II) (customized version of SELP)
W. Virginia	ELDA, but renamed West Virginia Test for English Language Learners (WESTELL) for use in the state
Wisconsin	ACCESS
Wyoming	ACCESS

SOURCE: http:// www.ncela.org; data confirmed by the state Title III directors.

TABLE 3-2 Tests Reviewed by the Panel

Test	States That Use the Test
ACCESS	Alabama, Delaware, DC, Georgia, Hawaii, Illinois, Kentucky, Maine, Mississippi, Missouri, New Hampshire, New Jersey, New Mexico, North Carolina, North Dakota, Oklahoma, Pennsylvania, Rhode Island, South Dakota, Vermont, Virginia, Wisconsin, Wyoming
CELDT	California
CELLA	Florida
ELDA	Arkansas, Iowa, Louisiana, Nebraska, South Carolina, Tennessee, West Virginia
LAS Links*	Colorado, Connecticut, Indiana, Maryland
NYSESLAT	New York
SELP*	Arizona, Washington
TELPAS	Texas

*Test is customized for each state so that it measures the state's English language proficiency content standards.

vided, and we did a cursory review of the procedures that were used, but we did not conduct a full-scale evaluation of each test. For example, we examined the technical manuals to confirm that reliability and validity information was reported, but we did not evaluate the procedures for obtaining reliability and validity information or the quality of the information reported. Doing the latter would have required that we first agree on the criteria for evaluating the tests and then thoroughly review the processes each used and the data each reported. Time and resources for this project were too limited to perform this type of review. The information we report about the tests is primarily descriptive and intended to support our charge of evaluating the extent to which the test results yield valid and comparable decisions across the states.

GENERAL SIMILARITIES AND DIFFERENCES AMONG THE STATE TESTS

In the most general sense, the new ELP tests have much in common, which is understandable since all of them were designed to meet the new requirements of NCLB. All assess ELP in the four broad domains specified by the legislation: listening, speaking, reading, and writing. All assess academic language as conceptualized and defined in the ELP content standards, are standards based (i.e., designed to evaluate the ELP standards set by the state), and are aligned with the language demands in the state's core academic content standards (discussed above).

In this section we discuss the similarities and differences among the tests with regard to their content standards, the grade bands (i.e., clusters of grades) covered by the tests, the item types, the scores reported, the criteria used to determine ELP, the methods used to set cut scores, and the reliability and validity of the tests.

English Language Proficiency Content Standards

When NCLB was enacted, one requirement was that states develop and/or adopt a set of ELP content standards to define the knowledge and skills that ELL students would be expected to master. Some states adopted the standards developed by an organization called Teachers of English to Speakers of Other Languages (TESOL), an association whose mission is to develop and maintain professional expertise in English language teaching and learning.[12] Other states created their own standards or made adjustments in the TESOL standards to meet their own needs. Articulation of the set of knowledge and skills that students should know and be able to do is the first step in designing a test, and it has a major impact on the nature of the test. Thus, while all of the tests measure ELP, they measure the skills of listening, speaking, reading, and writing in different ways.

Three of the tests that we reviewed (CELDT, NYSESLAT, and TELPAS) were developed specifically for a given state and thus are designed to measure that state's

[12]For information, see www.tesol.org [December 2010].

proficiency standards. Three tests (ACCESS, CELLA, and ELDA) were developed through one of the state consortia and so had to derive a strategy for dealing with differing state standards.

The strategy used for developing the ELDA standards provides an example of this process (American Institutes for Research, 2005). For this test, the ELP standards were defined through a synthesis of the standards used by the original states in the consortium.[13] The standards were initially merged by the test developer (American Institutes for Research) and were then refined by a consortium steering committee. The group agreed to common standards for each of the four domains. Some member states used these ELDA standards to guide the adoption of their own standards. Other member states reviewed their existing standards for alignment with the ELDA standards and made adjustments as needed. The result of this process was that all the states using the ELDA adopted similar ELP standards.

The WIDA Consortium used procedures similar to those used by ELDA for identifying the test standards, as did SELP and LAS Links. For instance, for SELP, the test framework was originally based on an analysis of ELP standards for six states (California, Delaware, Hawaii, Georgia, Missouri, and Texas) in conjunction with a review of the TESOL standards. Alignment studies were used to evaluate the correspondence of a particular state's standards with the test itself, and adjustments were made as needed (Pearson Education, 2009). Most states that administer the SELP or LAS Links use an augmented version, meaning that items are added to ensure that the test measures a state's standards and meets its specific needs.

Grade Bands

NCLB requires that ELP tests be available for students at all levels, from kindergarten through 12th grade, and so the assessments have different versions of the test for specific clusters of grades. As noted above, there are usually versions for the early elementary grades, later elementary grades, middle school, and high school, although the specific span of grades varies across tests:

- ELDA and ACCESS have versions intended for five grade bands: pre-K to kindergarten, grades 1 and 2, grades 3 through 5, grades 6 through 8, and grades 9 through 12.
- Texas has versions of the TELPAS for seven grade bands (K-1, 2, 3, 4-5, 6-7, 8-9, and 10-12).
- Washington and Arizona, which both use customized versions of the SELP, have versions of the test for different grade bands: both have versions for upper elementary (3-5), middle school (6-8), and high school (9-12); Washington uses a version for grades K-2, while Arizona has two versions

[13]Initially, 18 states participated in the consortium, and 6 had ELP standards in place.

of the test for these grade bands, one for kindergarten and one for grades 1-2.

All of the test programs have implemented vertical linking procedures to enable comparisons of performance across adjacent grade bands.

Item Types

Some tests use strictly multiple-choice questions (e.g., the CELLA); others use a combination of item types. For instance, the ACCESS uses multiple-choice questions for reading and listening and constructed-response questions for writing and speaking. The SELP and CELDT use a combination of multiple-choice and constructed-response (both short answer and extended answer) for each of the domains. The TELPAS uses classroom-based performance evaluation for all domains except reading.

Research has shown that performance on constructed-response and performance-based items is not entirely equivalent to performance on multiple-choice items. That is, students with the same level of writing skills might perform somewhat differently on the multiple-choice questions used by the CELLA than on the constructed-response questions used by the ACCESS, which primarily require expressive skills, or on the classroom performance-based items on the TELPAS. However, the different item formats generally measure related constructs and can usually be combined into a single scale (Ercikan et al., 1998).

Scores

Nearly all of the tests we reviewed report scores for each of the domains (listening, speaking, reading, and writing), an overall composite score summarizing performance in all four domains, and a comprehension score that is a composite of performance on the listening and reading tests. The NYSESLAT is an exception in that it reports two composite scores, one for listening and speaking and one for reading and writing. Some tests (ACCESS, CELLA, LAS Links, and Arizona's version of the SELP) also report an oral language score, which is derived from performance on the listening and speaking tests. In addition, the ACCESS test provides a score for literacy, based on combined performance on the reading and writing tests.

Although the tests all report some type of composite score, these composites are not consistently based on either equally or unequally weighted subscale scores. For instance, the CELDT and CELLA assign equal weights to the domain scores in determining the overall score. Other tests weight the domain scores differentially. The overall score on the TELPAS accords the most weight to the reading test (75 percent), the writing score is weighted by 15 percent, and the listening and speaking scores are weighted by 5 percent each. For ACCESS, the overall score weights reading and writing by 35 percent each and weights listening and speaking by 15 percent each. For the consortium-based tests (ACCESS, ELDA), the scores that are reported and any weights that are used are the same for all states using the test. For LAS Links and SELP, the test publisher offers a number of options for the states that use the tests. Thus, the overall score for Washington's version of the SELP may reflect a different weighting of the composite scores than the overall score for Arizona's version of the SELP.

This differential weighting reflects states' priorities with regard to which aspects of English proficiency in the four domains are acquired first and which domains are critical to succeeding in school. For instance, the technical guide to the TELPAS notes that listening and speaking are intentionally accorded less weight than reading and writing to ensure that students do not obtain a high overall score without acquiring the necessary skills in reading and writing. Young children usually acquire listening and speaking skills first (DeÁvila, 1997; Hakuta et al., 2000). Older students who have been schooled in academic subject matter in their native language can learn to read English text fairly quickly once they have studied the subject and learned basic English vocabulary, grammar, and structure. If a student is not literate in the native language and has had minimal or interrupted schooling or has not been taught the subject matter, learning to read and write will take more time, and these skills will be more difficult to master (Hakuta et al., 2000; Parker et al., 2009).

Weighting of language proficiency domains in ELP tests is important because it means that the skills represented by the overall scores differ from test to test. And we note that all of these weighting schemes simply refer to the weights applied in combining raw scores or scale scores for the four domains. Even when these nominal weights agree across different tests, the relative influence of different domains on the tests' composite scores may differ because the relative influences are also affected by the variances of the subtests for each domain.

English Language Proficiency Levels

Using the results from an ELP assessment, states are required to report the number of students who made progress in learning English (the first annual measurable achievement objective, or AMAO1) and the number who attained ELP (AMAO2) each year. In order to accomplish this, each of the test publishers has developed a number of categories of performance, referred to as "performance levels" (also referred to as "proficiency levels" or "achievement levels"). In order to determine the scores on the test that are considered to define the boundaries for each of the given performance levels, a standard-setting procedure must be used. Standard setting is a process for determining the minimum score (or "cut score") that a student must obtain on the test to be considered as having attained a given proficiency level. Standard setting is typically accomplished by using a set of trained participants who make judgments about how scores on a test relate to performance descriptors for each proficiency level. These judgments are used to set the cut score for each of the performance levels.

All ELP tests now being used have established performance levels and have used formal standard-setting methods to determine the cut scores for each performance level. They provide narrative descriptions of the knowledge and skills each performance level represents. These "performance level descriptions" characterize stages of language learning that can be used to determine test takers' instructional needs.

For accountability purposes, each state is required to determine a level of performance on its ELP test that is considered to be "English proficient" and to annually report to the DoEd the number of students who achieved this level (AMAO2). On this point it is important to distinguish between the *performance level on the test* that is designated as English proficient and the *process that a state uses to classify* a student as English proficient. The classification of a given student as English proficient may include criteria other than the student's score on the English proficiency test. The rules for this classification are discussed in greater detail in the next chapter. Here we are concerned only with how states determine the proficient level on the test; not surprisingly, this process varies from state to state. The definition of proficiency is determined differently in each state, using varying types of information, such as judgments of the standard setters; information external to the test, including the use of empirical analyses (e.g., analyses involving decision consistency between ELP and achievement tests or regression analyses of ELP test scores and academic assessment results); and judgments by policy makers or administrators. As such, the states have adopted different operational definitions of "English proficient" performance on their tests.

In some states, the definition is based on "conjunctive" rules, whereby students must meet all of a series of conditions. For example, California uses five performance levels to report performance on the CELDT: beginning, early intermediate, intermediate, early advanced, and advanced. To meet the standard of English proficient on the test, a student must have an overall score at the early advanced level, and all domain scores must be at the intermediate level or higher. In other states, the definition of English proficiency is based on compensatory rules. That is, high performance in one area or domain can compensate for lower performance in another. For example, California has an alternate definition that is based on compensatory rules: California students may also be judged to be English proficient if their overall score is at the high end of the intermediate level and there is other evidence of proficiency, such as scores on other tests, report card grades, and teacher evaluations.

The ELP tests differ in three important ways with regard to performance levels. First, the performance levels vary across the tests. The tests have different numbers of performance levels, different labels for them, and different descriptions for the skills they represent. This variation in performance levels is evident even among states that essentially use the same test (i.e., the states that use customized or augmented versions of SELP). Although there may be more similarities among the performance levels than is apparent at a surface level—for example, some use the same terminology to describe the skills they represent—there have been no qualitative or quantitative studies to evaluate the similarities and differences among the levels.

Second, the tests vary in the level of performance that is judged to be "English proficient" for meeting the accountability and reporting requirements of NCLB.

Here again, there may be variation even among states that use the same test. For instance, although the WIDA Consortium has adopted performance levels for the ACCESS, it is up to the state to determine the level that defines when a student is considered "English proficient." The same is true for the ELDA test developed by the LEP-SCASS.

Third, although all of the tests use a formal standard-setting procedure to set the cut scores, the standard-setting procedures differ across the tests. For instance, ACCESS, CELDT, CELLA, and LAS Links used the bookmark method;[14] ELDA used the bookmark method for all subtests except writing; NYSESLAT used an item mapping approach, similar to the bookmark approach, and SELP used the modified Angoff approach.[15] These different approaches can yield different results, as can the same approach used at different times. Research has also shown that the standard-setting results for the same test can vary depending on the particular set of judges that participate and the particular approach used (Impara and Plake, 1997; Jaeger, 1989; Kiplinger, 1996; Loomis, 2001; Musick, 2000; National Research Council, 2005; Texas Education Agency, 2002).

No studies have been done of the extent of differences among state performance levels. When tests use the same label for a performance level, such as "intermediate" or "proficient," one cannot simply assume that the same set of skills is represented or that a student who scores "proficient" on one test will also score "proficient" on another. Studies that do a crosswalk comparison of the performance levels used by different testing programs would help determine the extent of comparability. We describe these approaches later in the chapter.

Methods to Set Cut Scores Empirically

During our review, we learned that some states have conducted studies to empirically derive the level that they define as "English proficient." These states have explored methods for using performance on the content area achievement tests required by Title I of NCLB (the English language arts and mathematics achievement tests) as a criterion for helping them to determine the "English proficient" level. The method considers how ELL students perform on both the ELP and content area assessments, classifying them as proficient or not proficient on each. The method then seeks to identify the "English proficient" level on the ELP assessment that most consistently classifies students as proficient or not proficient on the English language arts and mathematics tests. The goal is to determine a cut score on the ELP test that maximizes the proportion of correct classifications on the ELP test in

---

[14]In the bookmark method, standard-setting panelists are asked to go through a specially constructed test booklet (arranged in order by the estimated difficulty of the items) and mark the most difficult item that a minimally proficient (or advanced) student would be likely to answer correctly; for details, see Mitzel et al. (2001).

[15]In the modified-Angoff method, standard-setting panelists are asked to estimate the percentage of minimally proficient (or advanced) students who would be expected to answer each item correctly; for details, see Angoff (1971).

relation to both content tests. The cut score derived from this empirical analysis is then taken into consideration by a panel of judges when they set the cut score that defines proficient performance on the ELP test. This empirically based method has been used to set ELP cut scores for 12 states that use the ACCESS as well as several non-WIDA states (Cook et al., 2009).

Reliability and Validity

Detailed information about the technical qualities of the tests is available in their technical reports. For our review, we examined the reports to determine the type of information that was available, but we did not evaluate the quality of that information. For instance, we reviewed the technical information available about each test to determine if it included the appropriate kinds of analyses to examine score reliability, but we did not evaluate the methods used to estimate reliability or the adequacy of the reliability estimates.

The testing programs generally report measures of internal consistency for the tests that are based on multiple-choice questions, and they provide measures of interrater agreement for tests that use constructed-response questions and are scored by humans. Most of the tests also provide an analysis of classification consistency, which examines the extent to which students are accurately classified into the various performance levels.

Most of the testing programs have conducted studies to evaluate the fairness of their items and identify any items that are potentially biased. These studies usually entail reviews of the items by expert panels, although a few of the programs have conducted analyses of differential item functioning.

The programs do not report an extensive amount of validity evidence. Content-related validity evidence consists primarily of alignment studies. This work involves comparison of the items (or test blueprint) and the ELP content standards to evaluate the extent to which the items measure the intended content and skills. Construct-related validity evidence typically consists of correlations between each item and the total test score (i.e., point-biserial correlations) and intercorrelations among the subtest scores. A few of the programs have conducted factor analyses to verify the factor structure of the assessment. Only two testing programs provided evidence of criterion-related validity. Analysts carried out a study of ACCESS in which performance on the test was compared with a priori proficiency categorizations of students who participated in the field tests (MacGregor et al., 2009). In another study, analysts compared students' performance on ELDA with teachers' ratings of students' English language proficiency (Lara et al., 2007). Wolf et al. (2008) reported that a cut score validation study was conducted for the CELDT.

For several of the tests, there is also evidence of the extent of correspondence between the scores and performance on another ELP test. For example, data are available on the correspondence between ACCESS scores and the New IDEA Proficiency Test (New-IPT) (Kenyon, 2006a), the Language Assessment Scales (LAS) (Kenyon, 2006b), the Maculaitis Assessment of Competencies Test of English Language Proficiency (MAC II) (Kenyon, 2006c), and the Language Proficiency Test Series (LPTS) (Kenyon, 2006d). Data are also available on the relationships between scores on the ELDA and scores on the New-IPT and the LAS, as well as on the correspondence between scores on LAS Links and the LAS. For some of the tests (NYSESLAT, TELPAS, ACCESS), data are available on the relationships between performance on the proficiency test and the state's English language arts test.

Technical Quality and Comparability of the Tests

Our review of eight ELP tests covered the information available in their technical manuals and supplementary materials with regard to test development, setting standards, deriving and reporting scores, and determining the reliability and validity of the scores. For this set of tests, we found evidence that the assessments have been developed according to accepted measurement practices. Each of the testing programs documented its efforts to evaluate the extent to which the test scores are valid for the purpose of measuring students' language proficiency in English.

NCLB set requirements for the tests, and as a result, pushed forward efforts to standardize certain aspects of these assessments. To meet the legislated requirements, the new tests must have a number of common features, and we found evidence of these features in all eight tests that we reviewed. The tests are all standards-based. They all measure some operationalized conceptualization of academic language, in addition to social/conversational language, in four broad domains and report scores for each of these domains, as well as a comprehension score and one or more composite scores. They all summarize performance using proficiency or performance levels, and states have established methods of looking at overall and domain scores in order to determine their respective definitions of English language proficiency. The tests also have versions available for students in kindergarten through 12th grade, with linkages to enable measurement of growth across adjacent grade bands. These common features provide the foundation for a certain degree of comparability across the tests.

Nevertheless, there are a number of ways in which tests can differ even though they meet the requirements set by NCLB, and we found evidence of these differences in the eight tests that we reviewed. They differ in many important ways that are likely to affect the comparability of the results that are reported. For instance, we found evidence that the tests we examined differed in content coverage, the types of questions used, test length, and timing of administration. Other aspects of the tests, such as the theory about academic language that underlies the questions, the difficulty of the questions, and measurement accuracy at each score point also can affect their equivalence. These differences mean that we cannot simply assume that a student who scores at the intermediate or proficient level on one state's ELP test will score at the intermediate or proficient level on another. Evaluating the extent to which they are comparable requires empirical analyses that may involve quantitative or qualitative approaches.

A quantitative approach for evaluating the equivalency of different assessments and putting the results on the same scale is referred to as "linking" (or "scaling"). Linking is a statistical procedure that allows one to determine the score on one test that is essentially equivalent to a score on another test (see Holland and Dorans, 2006; Johnson and Owen, 1998; Linn, 1993; Mislevy, 1992; National Research Council, 1999a, 1999b). These analyses are not easy to conduct, in part because of the data that must be collected. There are three types of linking procedures—equating, scale aligning, and predicting. The procedures range from strong to weak in terms of the assumptions they require and the inferences they permit. The strongest type of linking is equating. Equating is possible when the two assessments are designed according to the exact same specifications. That is, the assessments are matched in terms of content coverage, difficulty, type of questions used, test length, and measurement accuracy at each score point (Haertel and Linn, 1996; Holland and Dorans, 2006; Linn, 1993; Mislevy, 1992; National Research Council, 1999a, 1999b). To enable equating, the two tests must be given in a way that allows one to establish the linking function between the two tests, such as by randomly assigning students to take one or the other test (a randomized groups equating design), or having the same students take both tests or a set of items that are common across both tests (a common-items equating design). Other equating designs are also possible, but the randomized group and common-items equating designs provide the strongest basis for equating.

When the scores on two tests are equated, they are considered interchangeable. For the states' ELP tests, equating is not possible because the basic requirements for this kind of linkage have not been met (i.e., matching assessments in terms of content coverage, difficulty, type of questions used, test length, and measurement accuracy at each score point). Thus, we cannot say that the test results are comparable from state to state under the strictest definition of comparability.

It may be possible to link the results of different ELP tests using the two procedures with less stringent assumptions, scale aligning and predicting. Scale aligning is conducted when the tests being linked measure different constructs, or they measure similar constructs but with different test specifications (Holland and Dorans, 2006, p. 190). The goal of the predicting linking procedure is to predict an examinee's score on one test from some other information about that examinee (i.e., a score on another test, scores from several other tests, and possibly demographic or other information) (Holland and Dorans, 2006, p. 188). Both procedures require a linking design like those used for equating—randomized groups, same tests for groups, or common items. Less rigorous, nonequivalent groups designs are also possible, but just as with equating, they provide a weaker basis for developing the linking function, and the inferences they permit are limited.

It may also be possible to compare the different ELP tests using a more qualitative approach, often referred to as a "crosswalk review." A crosswalk evaluation is a systematic judgment comparison of key aspects of tests, including the content standards it is intended to measure, how it measures these standards, how item responses are aggregated to summary scores, and other key elements such as the performance levels.[16] In the present context, the analysis would need to focus on the levels set by the state to define when a student is "English proficient." This approach would compare the performance levels in terms of what students are expected to know and be able to do in order to be considered "English proficient" to evaluate the extent to which the states require similar skills. The approach might compare the performance levels with other sacross the states. Or it might involve determining *a priori* a definition of English proficiency and evaluating each state's performance levels in relation to this definition. The *a priori* definition might be determined by the DoEd or through the use of an independent expert panel.

To date, no qualitative crosswalk studies or statistical linking studies have been conducted for any of the ELP assessments we reviewed. "Bridging" studies have been done that predicted performance on the ACCESS from performance on other ELP tests (the studies by Kenyon, 2006a-2006d, mentioned earlier), but these studies were restricted to the kinds of assessments in place prior to NCLB (e.g., IPT, LAS, LPTS, and MAC II). It is important to point out that this situation is not unique to the ELP tests. The content standards, tests, and performance standards that states use for other aspects of NCLB (e.g., the reading and mathematics achievement tests) also vary from state to state, and scores are not comparable across states. Furthermore, it is important to note that the ELP tests were not designed from the outset to yield comparable results across states. The development effort would likely have taken a much different focus had cross-state comparability been the original intent. It is always difficult to attach a new use to test results when the test has not been designed from the outset for that purpose.

CONCLUSION 3-1 Although the English language proficiency assessments that we reviewed share common features and many states use the same test, the level of performance that defines when a student is considered to be "English proficient" is set by each state. There is no empirical evidence that has been collected to evaluate the comparability of these levels across the states.

In closing, however, we point out that results from the ELP test are not the sole basis for decisions to classify ELL students. Even if the ELP tests were linked and their scores placed on the same scale, there are still differences among the states in their procedures and criteria for classifying students. We take up these issues in the next chapter.

[16] Crosswalk analyses are sometimes used for alignment studies to evaluate the extent to which test items are aligned with content standards (see, e.g., http://www.adultedcontentstandards.ed.gov/docs/fieldResources/writing/Using%20Crosswalks%20for%20Alignment%20Notes.doc [December 2010]). Crosswalk analyses have also been conducted in a variety of other settings (see, e.g., http://www.calpro-online.org/eric/webliog.asp?tbl=webliog&ID=24 [December 2010]).

4

State Procedures for Identifying and Classifying English Language Learners

Although the Elementary and Secondary Education Act (ESEA) provides an official definition of a limited English proficiency (LEP) student, or English language learner (ELL) student, the act leaves it to states to operationalize the definition and to determine procedures for identifying students in need of Title III services. Figure 4-1 provides an overview of the classification and reclassification procedures.

Every state has an initial identification process whereby it identifies the pool of linguistic minority students, assesses their level of English language proficiency (ELP) using either a brief ELP assessment (usually called a "screener" or a "placement test") or a full-scale proficiency assessment, and determines which linguistic minority students are English language learners and therefore in need of Title III services. All states also have a process by which they annually assess ELL students' progress in learning English, determine when they no longer need these services, and procedures for reclassifying students as former English language learners.

Each state has developed its own approach, so the criteria for classification into and exit from ELL status, and the specialized services associated with it, vary across states. In addition, some states permit local control with respect to ELL classification and reclassification: the state sets forth general guidelines for ELL classification and exit criteria but allows local school districts to determine some or all of the criteria and performance standards for ELL classification and Title III services. Thus, in these states, the criteria can also vary from district to district.

The criteria that states use for identifying students as ELL and as in need of Title III services ultimately determine the numbers that they report to the U.S. Department of Education (DoEd). Some states have relatively stringent entry criteria and relatively lenient exit criteria, which means they are providing Title III services only for students most in need. Other states have more lenient entry criteria and more

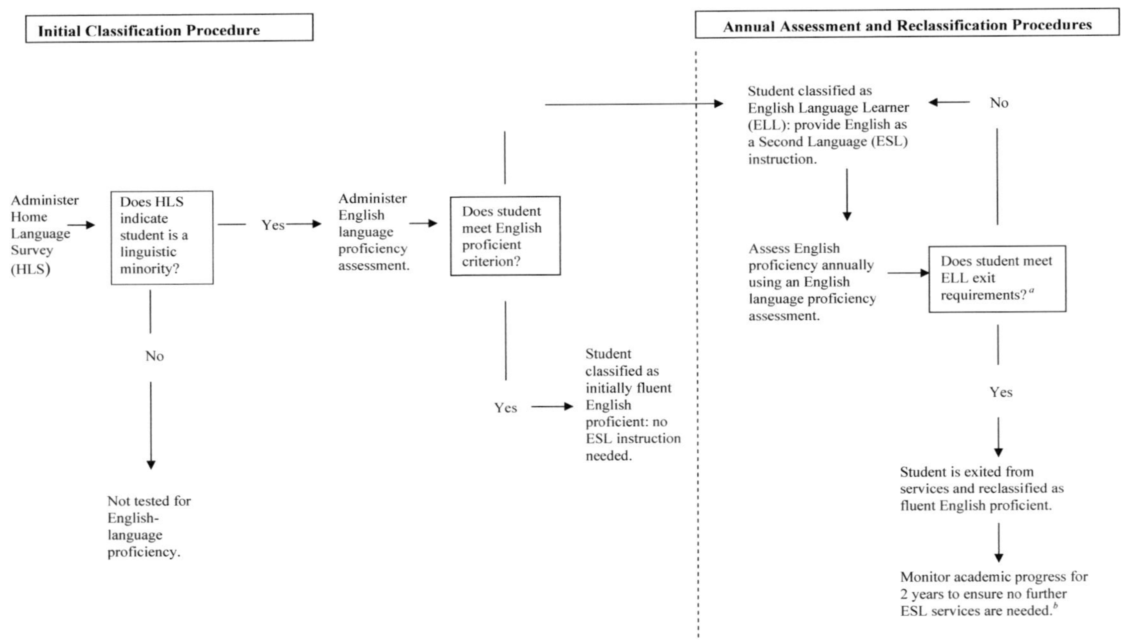

FIGURE 4-1 ELL classification and reclassification procedures.

[a]Some states use multiple exit criteria (e.g., a proficiency test, an academic achievement test, teacher's judgment, local assessments, parental input, language team review, etc.). They may be administered or monitored simultaneously or meeting one criterion may be required to sequentially "trigger" administration or review of others.

[b]Students who exit ESL must be followed for 2 years per federal law. On rare occasions, reclassified fluent English proficient students are judged to have been exited prematurely and are returned to ELL status to receive needed linguistic and academic services.

stringent exit criteria, which means they are providing services to many students with English language difficulties and retaining them in the classification until they are ready to function without specialized language and instructional support services. If Title III funding is going to be based on the counts provided by the states, it is important to understand the policies, processes, and practices that lead to these counts and the ways that the policies, processes, and practices differ across the states.

In this chapter, we compare the processes used by the states to classify students as ELL and therefore eligible for Title III services. After first commenting on the panel's approach to obtaining the information, we discuss states' procedures for initially classifying a student as an ELL. We then discuss states' procedures for reclassifying students as "formerly English language learners" and exiting them from the ELL category and its attendant specialized services. In the final section of the chapter we discuss the reporting mechanisms under which the data on ELL students are gathered, assembled, forwarded, and maintained and the effects of those mechanisms.

The committee relied on existing sources for information about state policies, practices, and criteria. The sources included several recent large-scale efforts to gather information on states' procedures for identifying ELL students:

- extensive information by Bailey and Kelly (2010) on home language surveys;
- data from Wolf et al. (2008) on state (including the District of Columbia) policies, procedures, and criteria for the 2006-2007 school year;
- an in-depth study by Ragan and Lesaux (2006) of the procedures in place during the 2004-2005 school year in 10 states and 10 school districts with high enrollments of ELL students;[1] and
- a study by Porta and Vega (2007) about states' procedures and their ELP tests.

These studies provided a snapshot of policies and practices prior to 2008-2009. For information about policies, procedures, and criteria in place in 2008-2009, the panel held focused reviews and discussions with officials in seven states: California, Colorado, New York, North Carolina, South Carolina, Texas, and Washington, all of which have high ELL student enrollments. In addition, we conducted a survey of state Title III administrators to update the information about the assessments their states use.

---

[1]The states of California, Texas, Florida, New York, Arizona, Illinois, Colorado, New Mexico, Georgia, and New Jersey (listed in order by size of ELL student enrollment) and the districts of Los Angeles, New York City, Dade County, Chicago, Houston, Santa Ana, San Diego, Long Beach (CA), Clark County (NV), and Broward County (FL).

INITIAL CLASSIFICATION OF STUDENTS

Home Language Surveys

The first step toward students' initial classification as an ELL student, as shown on the left-hand side of Figure 4-1, is administration of a brief questionnaire referred to as the home language survey (HLS). Use of an HLS to identify linguistic minority students originated with the remedies implemented in 1975 after the Supreme Court decision in the *Lau v. Nichols* discrimination case.[2] Although school systems are not required to use an HLS under federal law, Title III does require states to identify students in need of language support services, defining such individuals in terms of coming from "an environment where a language other than English has had a significant impact on the individual's level of English language proficiency" (Bailey and Kelly, 2010, p. 1). In the absence of federal guidance, all states have recommended, and most have required, the use of an HLS as the first step in identifying students that are potentially in need of language support services.

States and local school districts vary with regard to administering the HLS, beginning with when the questionnaire is administered. In some states, it is routinely administered to all students at the time of enrollment. Typically, the school official that handles new school enrollments asks the questions of the parent(s) or other adult guardian enrolling the child. In other states, the questionnaire is administered only to students who are referred for ELL services: referrals are typically made by parents and teachers. The questionnaires also differ with respect to the phrasing and content of the questions asked and with respect to the state regulations for implementation and interpretation of responses for further screening and assessment (Bailey and Kelly, 2010, p. 4). Further, some states are "local control states" and allow the school district to have final say over the questions used on the HLS.

We asked officials in the seven states that we studied about the questions asked on their HLS and received the information below.

California:

California is a local control state. The HLS includes, but is not limited to, the questions listed below. The local school district may add questions.

- Which language did your child learn when he/she first began to talk?
- Which language does your child most frequently speak at home?
- Which language do you (the parents or guardians) most frequently use when speaking with your child?
- Which language is most often spoken by adults in the home? (parents, guardians, grandparents, or any other adults)

---

[2]See http://scholar.google.com/scholar_case?case=7491158078497524278&q=lau+v.+nichols&hl=en&as_sdt=80000000000002&as_vis=1; http://www.pbs.org/beyondbrown/brownpdfs/launichols.pdf [November 2010].

Colorado:
- Is a language other than English used in the home?
- Was the student's first language other than English?
- Does the student speak a language other than English?

New York:
- What language does the child understand?
- What language does the child speak most often?
- What language does the child read best?
- What language does the child write best?

North Carolina:
- What is the first language the student learned to speak?
- What language does the student speak most often?
- What language is spoken in the home?

South Carolina:
- What is the language that your child first learned?
- What language does the student speak most often?
- What language is most often spoken in the home?

Texas:
- What language is spoken in the home most of the time?
- What language does the child speak most of the time?

Washington:
- Is a language other than English spoken in the home?
- Did your child first speak a language other than English?
- What language did your child first speak?

These examples illustrate the variety of questions used to initially identify students for ELL services in just a few states. (For a more comprehensive listing of state questions, see Bailey and Kelly, 2010.)

In their analysis of the questions, Bailey and Kelly (2010) classify the differences along several dimensions. Some questions focus on the first or native language of the child; other questions focus on where the other language is spoken or what languages other than English are spoken; still others focus on the frequency with which the student speaks English, equating language dominance with proficiency.

State practices for implementing an HLS also vary. In some states, the questions are standardized through a stateside mandated form. In states with local control, each school district determines the questions, often through use of sample question forms provided by the state that local school districts are encouraged to adopt. For instance, in Colorado and California, which have local control, the questions are recommended, but each district determines the exact questions to ask.

States also differ with respect to follow-up to the HLS. When any response to an HLS suggests that the student may not be a native English speaker, some follow-up action is taken, generally within a prescribed time period of 30 days (Title III law, Part C-General Provisions, SEC. 3302(a)). Typically, the next step is to administer an initial assessment to evaluate the student's level of English proficiency, although in some states, a parent or student interview may be conducted before deciding whether or not to administer a proficiency assessment.

The validity of classifications based on the HLS has been questioned. For instance, Abedi (2008) noted that parents may not respond accurately because they misunderstand the questions, are worried about providing information that could lead to questions about their citizenship status, or are concerned that the ELL classification will lead to restricted educational opportunities for their child. These factors are likely to be more of an issue with recent immigrants or undocumented immigrants and in states with high populations of either group.

In summary, most states use an HLS as the first step in the ELL classification process—the initial identification of students as linguistic minority and therefore potentially English language learners. However, the number and content of the questions, the administration procedures, and decision rules about the results vary from state to state and, in some states, from district to district.

Initial English Language Proficiency Assessments

The initial assessment of a student's English language proficiency usually involves administering a test. States use a variety of tests for this purpose, which tends to affect the comparability of the state data. Some administer the full state ELP test that is used for federal annual accountability reporting. Other states use a brief proficiency assessment, often called a "placement test," or a "screener." The objective of these tests is to further determine the student's level of proficiency in four language domains (speaking, listening, reading, and writing) after the initial HLS inquiry of language environment, preference, and use (typically in speaking) has signaled linguistic minority status.

Table 4-1 shows the tests that each state uses for initial classification of students in need of Title III services.

A majority, 27 states, use a screener test. Of these 27, 18 use one of the screener tests developed by the World-Class Instructional Design and Assessment (WID Consortium (the W-APT or the MODEL), 3 use the LAS Links Placement test, 4 use their own screener, 1 uses the LAB-R, and 1 uses the Woodcock Munoz Language Survey.[3] Four states use their ELP test for the initial proficiency assessment (Alaska, Arizona, California, and Florida), while two states (Connecticut and Nevada) allow

[3]For the full names of the tests, see Table 4-1.

TABLE 4-1 Tests Used by the States for Initial Classification of English Language Learners for the 2009-2010 School Year

State	Test Used	Type of Test
Alabama	W-APT	WIDA Screener/placement test
Alaska	New IDEA Proficiency Test	ELP test
Arizona	AZELLA	ELP test (customized form of the SELP)
Arkansas	District chosen (LAS II or MAC II)	Combination
California	CELDT	ELP test
Colorado	CELA Placement test	Screener/placement test
Connecticut	LAS Links Placement Test, LAS, or any ELP test	Screener/placement test or ELP test
Delaware	W-APT or MODEL	WIDA Screener/placement test
DC	W-APT	WIDA Screener/placement test
Florida	CELLA screener, LAS, or other test chosen by the district	Combination
Georgia	W-APT	WIDA Screener/placement test
Hawaii	LAS Links Placement Test	Screener/placement test
Idaho	Idaho English Language Assessment (IELA)	Screener/placement test
Illinois	W-APT, MODEL	WIDA Screener/placement test
Indiana	LAS Links Placement Test	Screener/placement test
Iowa	LAS, IPT (district chosen)	
Kansas	KELPA, KEOPA-P, IPT, LAS, LAS Links, or LPTS (district chosen)	Combination
Kentucky	W-APT	WIDA Screener/placement test
Louisiana	District chosen	Combination
Maine	W-APT or MODEL	WIDA Screener/placement test
Maryland	LAS Links Placement Test	Screener/placement test
Massachusetts	District chosen	Combination
Michigan	ELPA Initial Screening	Screener/placement test
Minnesota	District chosen	
Mississippi	W-APT	WIDA Screener/placement test
Missouri	W-APT	WIDA Screener/placement test
Montana	District chosen	Combination
Nebraska	District chosen	Combination
Nevada	Pre-LAS or LAS Links	Pre-LAS is a screener test; LAS Links is an ELP test
New Hampshire	W-APT	WIDA Screener/placement test
New Jersey	District chosen	Combination
New Mexico	W-APT	WIDA Screener/placement test
New York	Language Assessment Battery-Revised (LAB-R)	Screener/placement test
N. Carolina	W-APT or MODEL	WIDA Screener/placement test
N. Dakota	W-APT	WIDA Screener/placement test
Ohio	District chosen	Combination
Oklahoma	W-APT	WIDA Screener/placement test

continued

TABLE 4-1 Continued

State	Test Used	Type of Test
Oregon	District chosen	
Pennsylvania	W-APT	WIDA Screener/placement test
Rhode Island	W-APT	WIDA Screener/placement test
S. Carolina	District chosen (Woodcock Munoz Language Survey, LAS, IPT)	Combination
S. Dakota	W-APT	WIDA Screener/placement test
Tennessee	State developed test designed to be aligned with the ELDA	Screener/placement test
Texas	District chosen	Combination
Utah	District chosen	Combination
Vermont	W-APT	WIDA Screener/placement test
Virginia	W-APT, or district chosen	WIDA Screener/placement test
Washington	WLPT-II Placement	WLPT Screener/placement test (customized version of the SELP)
W. Virginia	Woodcock Munoz Language Survey	Screener/placement test
Wisconsin	W-APT	WIDA Screener/placement test
Wyoming	District chosen (but all used W-APT)	WIDA Screener/placement test

SOURCE: http://www.ncela.org; data confirmed by state Title III director.

districts to choose between the state ELP test or the screener. In addition, 17 states allow districts to select the language proficiency assessment used for initial classification, though they generally provide a list of tests from which the district can select.

Examples of Initial Classification Procedures

Our discussions with Title III officials of the seven states we studied helped to clarify the steps and decisions involved in the initial classification process. The information is summarized below.

California Students who are identified as having a primary language other than English based on the HLS, must be assessed on the California English Language Development Test (CELDT). The CELDT is the designated state test of English language proficiency. Therefore, pupils must achieve the English proficiency level on the CELDT to be classified as Initially Fluent English Proficient. To achieve the English proficient level on the CELDT, pupils at grades 2 through 12 must have an overall score of Early Advanced or above and all four domains (listening, speaking, reading, and writing) at Intermediate or above. Pupils in Kindergarten and grade 1 (K-1) must have an overall score of Early Advanced or above and listening and speaking domains at Intermediate or above. Students who do not score at this level are classified as ELL students.

Colorado Students who are initially identified through the HLS are given a screener test, called the Colorado English Language Assessment (CELA) placement test. The screener assesses all four modalities (reading, writing, speaking, listening). A score of "Approaching Proficiency" or below indicates that ELL services are needed.

New York A response other than English to any of the questions on the HLS triggers an informal interview conducted in the native language and in English. If the student speaks little or no English, the student is assessed with the Language Assessment Battery-Revised (LAB-R), the state's screener test. Those who score below the proficient level are classified as ELL students.

North Carolina Staff of the state's English as a Second Language Program review the responses on the HLS, interview the parent or guardian as necessary, or observe the student to determine the home language. If it is determined that a student's home language is other than English, the state's screener test, the WIDA-ACCESS Placement Test (W-APT) is administered. For grades 1-12, those who score a composite of less than 5.0 or less than 5.0 on any of the four domains are identified as LEP students. For the first semester kindergarten W-APT, those who score less than 27 on listening and speaking are identified as LEP students. For the second semester kindergarten W-APT, those who score less than 27 on listening and speaking, less than 14 on reading, or less than 17 on writing are identified as LEP students.

South Carolina If the response to any of the questions on the HLS is a language other than English, the student is further assessed. The state currently allows districts to choose from among the LAS, IPT, or Woodcock-Muñoz assessments. The state plans to adopt the ELDA placement test as an additional screener when the test becomes available.

Texas A response other than English to either of the questions on the HLS triggers additional assessment. The local district is allowed to determine the assessment used for initial identification, provided it is one of the tests approved by the state education agency.[4] For pre-K through 1st grade, districts are to use a test of oral language proficiency. For grades 2 and higher, students are given a norm-referenced achievement test in reading and language arts (such as the Iowa Test of Basic Skills). If the student scores below the 40th percentile, the case is sent to a review committee.

Washington A response of "yes" to the second question on the HLS ("Did your child first speak a language other than English?) triggers additional evaluation with the state screener test, which is based on the SELP and called the WLPT-II (Washington Language Proficiency Test-II). Scoring at a level of 3 or below indicates

---

[4]For a list of agency-approved tests for the 2010-2011 school year, see http://ritter.tea.state.tx.us/curriculum/biling/ListofApprovedTests2010_2011.pdf [December 2010].

that the student is in need of ELL services. When responses to the HLS reveal that a language other than English is spoken in the home (first question) but the child speaks only English (second question), districts are required to follow up with the parent or guardian to ensure that the questions were clearly understood. ELL students are generally identified through the HLS, but teachers can also make referrals on the basis of their classroom observations.

In summary, states use either the ELP test or a screener or placement test as the second step in the ELL classification process to determine which linguistic minority students are English language learners. However, the assessments and criteria used for initially classifying students as English language learners vary from state to state. In addition in some states, a variety of other criteria may be considered in the initial classification decisions. And in states that permit local control, the assessments and the criteria used for initially classifying students as English language learners may vary from district to district.

CONCLUSION 4-1 Because of the differing state policies, practices, and criteria for initially identifying students as linguistic minority and for classifying them as an English language learner (ELL), individuals who are classified as ELL students in one state may not be classified as ELL students in another. In states that permit local control, students classified as ELL in one district may not be classified as ELL in another district in that state.

RECLASSIFICATION OF ELL STUDENTS

Criteria Considered

Each year, all ELL students must be reassessed to evaluate their progress in learning English. Typically, students are given an ELP test, and those who score at the level that the state has defined as "English proficient" may be considered eligible for reclassification as a "former ELL" student. The proficiency level that is the threshold for reclassification is determined by the state. This step of the process is shown on the right-hand side of Figure 4-1 (above).

In addition to scoring at the "English proficient" level on the ELP test, states consider a variety of other criteria in the reclassification process, such as performance on content area tests; input from school personnel; input from the parents or guardians; and other such measures as student grades, portfolios of the student's work, student interviews, and evaluations of classroom performance. In local-control states, these criteria may differ by district within a given state. States use different combinations of these criteria, some using only the ELP test and some using as many as six different types of criteria.

According to the information gathered by Wolf and colleagues (2008) for the 2006-2007 school year, 12 states consider only the ELP test score in reclassification

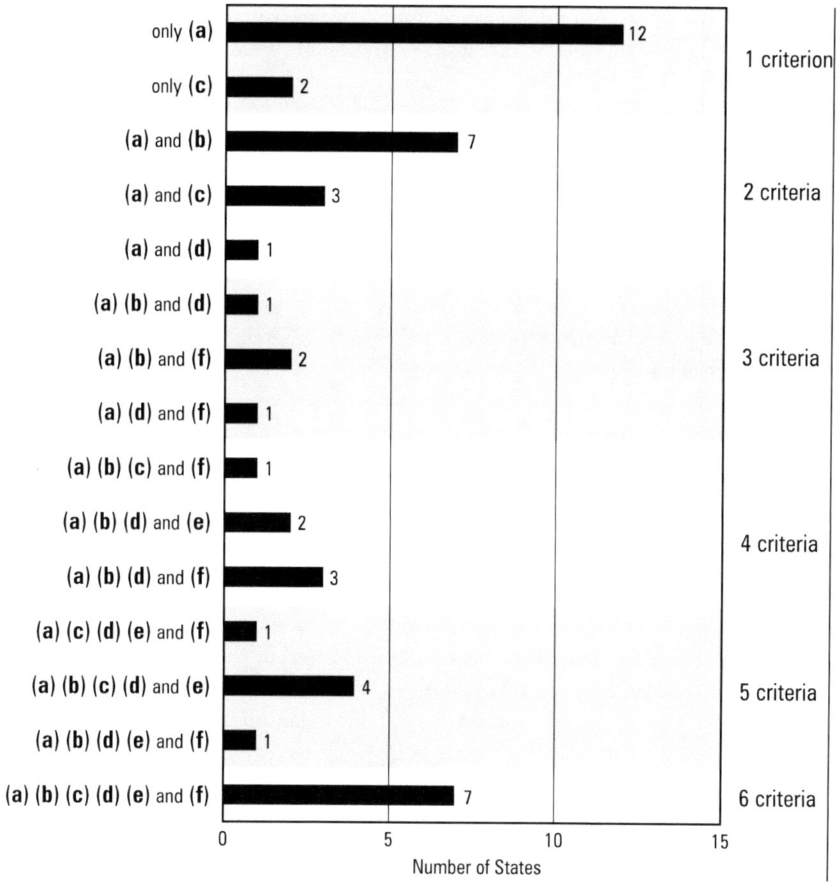

FIGURE 4-2 Number of criteria used by states for ELL reclassification.
NOTES: (a) Proficiency test scores, (b) content scores, (c) district-established criteria, (d) school personnel input, (e) parent or guardian input, (f) other.
SOURCE: Wolf et al. (2008). Figure from *Issues in Assessing English Language Learners: English Language Proficiency Measures and Accommodation Uses—Practice Review*. National Center for Research on Evaluation, Standards, and Student Testing, University of California, Los Angeles. Copyright ©2008. The Regents of the University of California and supported under the Institute of Education Sciences, U.S. Department of Education. Available: http://www.cse.ucla.edu/products/reports/R732.pdf. Reprinted with permission.

decisions, and 11 consider the ELP test and a second type of criterion—content-area achievement scores (7 states), district-level criteria (3 states), and other criteria (1 state): see Figure 4-2.

Of the remaining states, 26 indicated that they use the ELP test and two other kinds of criteria, and 2 indicated that the criteria are established by districts and thus vary across the state. It is important to note here that the survey by Wolf and colleagues was conducted prior to the implementation of recent federal regulations and interpretations. The results provide a general sense of the variability across the states, but they are somewhat out of date. Most importantly, current federal regulations require that the results of the ELP test be used in reclassification decisions (Federal Register, 2008, October 17).

Examples of Reclassification Procedures

Our conversations with Title III officials at the seven states we studied helps to clarify the steps and decisions involved in the re-classification process. The information is summarized below.

California California *EC* 313(d) specifies four criteria that local education agencies (LEAs) at a minimum must use for students to be reclassified as fluent english proficient.[5] Based on *EC* Section 313(d), guidelines for the reclassification of English learners were approved by the State Board of Education (September 2002, updated September 2006, and modified in July 2010). In accordance with *EC* 313(d), the reclassification procedures developed by the California Department of Education use multiple criteria in determining whether to reclassify a pupil as proficient in English, including, but not limited to, all of the following:

(1) performance on the CELDT;
(2) comparison of performance in basic skills against an empirically established range of performance in basic skills (e.g., the California Standards Test for English-Language Arts [CST for ELA] or the California Modified Assessment for ELA [CMA for ELA])[6];
(3) teacher evaluation, including, but not limited to, a review of the pupil's curriculum mastery; and
(4) parental opinion and consultation.

Colorado A body of evidence approach is used for reclassification. An overall score of 5, which corresponds to an advanced level on the Colorado English Language Assessment (CELA), triggers a broader review of the body of evidence available about the student. To be eligible for reclassification, a student must attain scores of 4.5 or higher on the subtests of the CELA and should score at the partially proficient level on the state's English language arts assessment (the Colorado Student Assessment Program, CSAP). Also considered are district review committee evaluations,

---

[5]For details, see http://www.cde.ca.gov/ta/tg/el/documents/celdt09astpkt1.pdf [November 2010].

[6]On the CST for ELA or the CMA for ELA, to be considered for possible reclassification, a pupil should score in a range from the bottom to the midpoint of the scale score range that defines the "Basic" performance level or higher.

language samples, observations, district-content specific tests, logs or journals, and other district-determined criteria.

New York Reclassification is based on results from the New York State English as a Second Language Achievement Test (NYSESLAT), the state's ELP test. Students who score below the proficient level continue to receive ELL services. Students who score at the proficient level enter the general education program. No other criteria are considered.

North Carolina Students are judged to have attained English language proficiency and so exit from LEP status if they have an overall composite score of 4.8, with at least a 4.0 on the reading subtest and at least a 4.0 on the writing subtest, on the state's proficiency test, ACCESS.

South Carolina To be reclassified, students must score at Level 5 on the English Language Development Assessment (ELDA), the state's ELP test. For grades 3-12, ELL students must pass with a composite score of 5 or higher. Because the ELDA inventories for grades K-2 do not measure academic English to the extent deemed necessary for ending ESL services, students at those levels cannot be reclassified.

Texas Students who receive ELL services cannot be reclassified before the end of first grade. Below grade 3, reclassification is based on achieving the mastery level on an agency-approved ELP assessment and an agency-approved English language arts norm-referenced test. For grades 3 and higher, the student must meet state standards for passing the state's English language arts test used for Title I of ESEA. Each district has a local language proficiency assessment committee that makes the final reclassification decision, based on a review of a variety of information available about the student. The review is a deliberative process. The committee gathers information from teachers and parents and assesses the needs of the "whole child." Passing the state English language arts assessment is a minimum requirement, but students may pass this assessment and still remain in the ELL classification if other information suggests that they still need ESL services. Reclassification (at any grade level) is also based on the results of a subjective teacher evaluation (which may include other assessments, anecdotal records, portfolios, etc.).

Washington Reclassification is based on performance on the Washington Language Proficiency Test (WLPT-II), the state's ELP test. Students must obtain an overall composite score of 4. Although Washington is a local-control state, the same reclassification standard is used by all school districts.

When a student meets the state (or local) criteria for reclassification, the student is formally "exited" from Title III services. This step in the process is shown on the right-most box on Figure 4-1 (above). These students are then reclassified as former ELL students, sometimes referred to as "reclassified fluent English proficient students" or "former English language learner" (the label given to these students differs across the states).

Under ESEA, former ELL students are to be monitored for 2 years after exiting ELL status to ensure that they continue to perform well without specialized support services. Students who exhibit academic challenges due to linguistic difficulties may be reevaluated and, in rare cases, may be classified as ELL students and provided Title III services.

In summary, all states use the results of an ELP test to identify students who are English proficient and eligible to be considered for reclassification. However, the proficiency test that is used and the level of performance that defines when a student is considered to be "English proficient" is determined by the state and thus varies from state to state and, in local-control states, may vary from district to district. In addition, some states use other criteria in making reclassification decisions, which may include performance on content-area tests; input from school personnel, parents, or guardians; student grades; portfolios of student work; student interviews; and evaluations of classroom performance. Use of these criteria and decision rules varies from state to state. In states that permit local control, these criteria and rules can vary from district to district within a state.

CONCLUSION 4-2 Because of the wide variety of state policies, practices, and criteria for reclassifying students as former English language learners, and thereby exiting them from Title III services, a given English language learner student may remain in the classification longer in one state than in another state. In local control states, similar variation may exist among districts within the same state.

DATA ON ELL STUDENTS REPORTED TO THE FEDERAL GOVERNMENT

Although the variety of procedures used by states for classifying and reclassifying is wide, it does not mean that the procedures are not well thought out and appropriate for each state. Each state has adopted a process that it believes suits its own needs. Each state is aware of the linguistic demands of its regular curriculum, and it has developed procedures that allow it to ensure that students receive the support and instruction they need until they have adequate English language skills to handle the curriculum.

The classification process in each state is comprehensive and multidimensional. It includes direct assessments of students' proficiency in English, along with other information about the student's proficiency in English—such as classroom grades and performance, parental input, and teacher input. Thus, while there may be little cross-state comparability in the ways that the classification decisions are made, there is evidence that states make use of a variety of types of information in making classification decisions. However, our objective in reviewing these state practices was to determine whether there was sufficient commonality of policies, practices, and criteria to ensure that data on the ELL population from one state would have the same meaning as data from another state if used in an allocation formula.

> **CONCLUSION 4-3 Each state has developed its own policies, practices, and criteria for determining which students need Title III services. Although these policies, practices, and criteria are intended to address the needs of English language learner (ELL) students in each state, they vary from state to state. And in states that permit local control in making these decisions, they may also vary from district to district. Consequently, counts of ELL students that qualify for Title III funding are based on criteria that can vary across the states, and among districts in some local control states.**

The counts of ELL students start with schools and LEAs, which have the primary responsibility for collecting, maintaining, and reporting data on selected student populations they serve, including the number and other characteristics of ELL students they serve. LEAs report these numbers to their respective state education agency (SEA). The SEA aggregates the school and LEA data and provides the information as part of the Consolidated State Performance Report that each state submits to the DoEd.

The student-level data that is reported by schools and LEAs are defined in instructions to the states from the DoEd. The data on LEP (ELL) students consist of four types of counts that are relevant to this report:

1. the unduplicated count of all students in the state who meet the ESEA definition of LEP (see Chapter 1), which includes newly enrolled students and continually enrolled students whether or not they receive Title III services;
2. the unduplicated number of students who received services in Title III language instructional education programs;
3. the number of LEP students who were tested on an annual ELP assessment; and
4. for students who took the annual ELP test in (3), the number who scored at the level the state defines as "English proficient."

At the DoEd, these data are maintained in the Education Data Exchange Network (EDEN) system, the backbone of which is the EDEN Submission System, an electronic data system to which states submit data on over 100 data groups at the state, district, and local levels. (The EDEN system is the main source of a series of regular reports in EDFacts, a system that centralizes performance data from the states.) Our analysis is based on data for the 2006-2007 school year.[7]

---

[7]Although the data have been reported since the 2003-2004 school year, the instructions for the reports have been modified over the past few years, in part because of issues raised in a report from the U.S. Government Accountability Office (see Chapter 1). Similarly, the standard list of EDEN reports

We obtained data on all four types of counts described above, but we focused our analyses on two of the counts, which we selected for conceptual reasons. The first is the unduplicated count of all students in the state who meet the ESEA definition of LEP (1 above). We judged that this count represented the most all-inclusive ELL estimate available from the states and is defined by law. However, as discussed earlier in this chapter, the law leaves it to states to operationalize this definition, and states' policies, practices, and criteria affect their counts of students. And in local-control states, these policies, practices, and criteria may not be consistent across the state. We therefore decided to analyze a second count that we judged would provide a relatively objective measure across the states and would be relatively less susceptible to intrastate differences in local-control states: the number of students who were determined to be not proficient in English on a state's ELP test (derived from 3 and 4 above). We refer to this count as "tested, not proficient."

As explained in Chapter 3, all states are required to determine a level of performance on the ELP test that defines when a student is "English proficient." All districts within a state use the same test for this purpose, and the "English proficient" level is consistent throughout that state. When students meet this criterion, they are eligible for consideration for reclassification, although other criteria may come into play, which may differ both within a state and from state to state. So an ELL count based on those scoring proficient on the ELP test provides an estimate that is based on a criterion that is common across the state.[8]

For the most part, all of these data were available through the EDEN system and were provided to us by the DoEd. For the data that were not yet incorporated into the EDEN system, we obtained them through the Consolidated State Performance Reports (CSPRs), either from staff at the Department or from reports available online. The sources of the data used are indicated in the discussions below.

Counts of all ELL Students

Table 4-2 shows the numbers of ELL students in each state for the 2006-2007, 2007-2008, and 2008-2009 school years, listed in order by the average of these counts across the three school years. It also shows each state's share of the total U.S. population of ELL students.

Comparison of the data in Table 4-2 shows considerable year-to-year fluctuations in the absolute numbers of ELL students in some states. In Nevada, for example, which had the largest fluctuation, there was a decrease of about 23,000 ELL students, or about one-third, between 2006-2007 and 2007-2008 and a similar-sized increase between 2007-2008 and 2008-2009. Rhode Island also exhibited a fairly

---

and the formats of those reports have changed considerably over the years and have been the same for many reports only since the 2006-2007 school year. Thus, we obtained data from the U.S. Department of Education beginning with the 2006-2007 school year.

[8]We note, however, that reclassification policies, practices, and criteria define the population of students who take the proficiency test each year.

TABLE 4-2 Numbers and Shares of All ELL Students by State: School Years 2006-2007, 2007-2008, and 2008-2009

	2006-2007		2007-2008		2008-2009		
State	Number	Share*	Number	Share*	Number	Share*	3-Year Average
California	1,559,146.0*	36.4	1,553,091.0*	34.3	1,515,082.0	33.7	1,542,439.7
Texas	501,333.0	11.7	693,031.0	15.3	718,388.0	16.0	637,584.0
Florida	234,614.0	5.5	231,403.0	5.1	226,122.0	5.0	230,713.0
New York	196,153.0	4.6	210,359.0	4.6	183,952.0	4.1	196,821.3
Illinois	172,950.0	4.0	189,926.0	4.2	204,737.0	4.6	189,204.3
Arizona	152,753.0	3.6	149,721.0	3.3	125,636.0	2.8	142,703.3
North Carolina	87,745.0	2.0	127,449.0	2.8	113,823.0	2.5	109,672.3
Colorado	89,881.0*	2.1	85,323.0	1.9	88,907.0	2.0	88,037.0
Virginia	86,392.0	2.0	84,345.0	1.9	87,026.0	1.9	85,921.0
Washington	84,761.0	2.0	80,694.0	1.8	82,711.0	1.8	82,722.0
Georgia	74,132.0	1.7	79,987.0	1.8	80,890.0	1.8	78,336.3
Nevada	70,548.0	1.6	47,049.0	1.0	75,952.0	1.7	64,516.3
Oregon	61,914.0	1.4	62,111.0	1.4	63,011.0	1.4	62,345.3
Minnesota	63,858.0	1.5	61,229.0	1.4	61,486.0	1.4	62,191.0
Michigan	69,705.0	1.6	51,465.0	1.1	60,945.0	1.4	60,705.0
New Mexico	60,711.0	1.4	61,207.0	1.4	53,970.0	1.2	58,629.3
New Jersey	54,433.0*	1.3	54,503.0*	1.2	54,154.0	1.2	54,363.3
Massachusetts	54,071.0	1.3	55,730.0	1.2	49,073.0	1.1	52,958.0
Utah	48,399.0	1.1	46,770.0	1.0	44,470.0	1.0	46,546.3
Pennsylvania	45,431.0	1.1	46,357.0	1.0	47,672.0	1.1	46,486.7
Indiana	42,536.0	1.0	46,304.0	1.0	45,760.0	1.0	44,866.7
Wisconsin	41,312.0	1.0	43,790.0	1.0	47,866.0	1.1	44,322.7
Maryland	34,332.0*	0.8	40,421.0	0.9	40,051.0	0.9	38,268.0
Oklahoma	38,109.0	0.9	37,744.0	0.8	38,092.0	0.8	37,981.7
Ohio	29,240.0	0.7	35,038.0	0.8	36,376.0	0.8	33,551.3
Kansas	28,915.0	0.7	31,760.0	0.7	34,096.0	0.8	31,590.3
South Carolina	30,163.0	0.7	28,366.0	0.6	31,450.0	0.7	29,993.0
Connecticut	26,357.0	0.6	30,033.0	0.7	29,751.0	0.7	28,713.7
Arkansas	23,651.0	0.6	25,905.0	0.6	27,634.0	0.6	25,730.0
Tennessee	23,009.0	0.5	25,670.0	0.6	27,428.0	0.6	25,369.0
Alabama	18,358.0	0.4	20,943.0	0.5	19,523.0	0.4	19,608.0
Iowa	18,124.0	0.4	19,442.0	0.4	20,334.0	0.5	19,300.0
Missouri	22,365.0	0.5	19,053.0	0.4	16,338.0	0.4	19,252.0
Nebraska	18,190.0	0.4	19,128.0	0.4	18,394.0	0.4	18,570.7
Hawaii	15,660.0	0.4	18,681.0	0.4	18,564.0	0.4	17,635.0
Idaho	16,698.0	0.4	16,671.0	0.4	17,669.0	0.4	17,012.7
Alaska	20,761.0	0.5	16,823.0	0.4	12,030.0	0.3	16,538.0
Kentucky	10,816.0	0.3	12,896.0	0.3	14,589.0	0.3	12,767.0
Louisiana	8,629.0	0.2	11,474.0	0.3	12,497.0	0.3	10,866.7
Rhode Island	10,034.0	0.2	7,149.0	0.2	9,397.0*	0.2	8,860.0

continued

TABLE 4-2 Continued

State	2006-2007 Number	Share*	2007-2008 Number	Share*	2008-2009 Number	Share*	3-Year Average
Delaware	6,648.0	0.2	7,260.0	0.2	7,184.0	0.2	7,030.7
Montana	6,983.0	0.2	6,722.0	0.1	4,550.0	0.1	6,085.0
Mississippi	4,982.0	0.1	5,451.0	0.1	6,543.0	0.1	5,658.7
District of Columbia	4,717.0*	0.1	5,126.0	0.1	5,854.0	0.1	5,232.3
Maine	3,691.0	0.1	4,036.0	0.1	4,215.0	0.1	3,980.7
North Dakota	2,399.0	0.1	4,648.0	0.1	4,068.0	0.1	3,705.0
South Dakota	3,291.0	0.1	4,217.0	0.1	3,594.0	0.1	3,700.7
New Hampshire	3,149.0	0.1	3,201.0	0.1	4,076.0	0.1	3,475.3
Wyoming	3,006.0	0.1	2,395.0	0.1	2,277.0	0.1	2,559.3
West Virginia	2,248.0	0.1	2,336.0	0.1	1,618.0	0.0	2,067.3
Vermont	1,743.0	0.0	1,459.0	0.0	1,495.0	0.0	1,565.7
U.S.	4,289,046	100.0	4,525,892	100.0	4,499,072	100.0	4,438,033.3

NOTES: States are listed in order by the 3-year average of their reported numbers of ELL students.
*The shares represent each state's share of the total number of ELL students in the country.
SOURCE: Data from the U.S. Department of Education, Education Data Exchange Network, except counts noted with an asterisk, which were obtained from the Consolidated State Performance Reports.

large proportional swing, declining by about one-third, from about 10,000 to 7,100 students, between 2006-2007 and 2007-2008.

Yet although the absolute numbers appear to fluctuate, the states tend to rank order quite similarly across the years in terms of their percentage share of the total ELL population in the country. For instance, the 11 states with the highest percentage shares of ELL students are the same across all 3 years. Approximately 75 percent of the nation's ELL students in the country reside in these 11 states.

Another way to think about the numbers of students served by each state is in relation to the total population of school-age children in the state. In effect, this percentage reflects the burden placed on the state: the percentage of its school-age population that needs Title III services. Table 4-3 shows each state's count of all ELL students as the percentage of the total number of K-12 students enrolled in the state's public schools,[9] which we refer to as the state ELL rate.

Comparison of these percentages across the 3 years shows that they tend to be

[9]Data on the total number of K-12 students enrolled in the state's public schools are from the Common Core of Data for the 2006-2007, 2007-2008, and 2008-2009 school years: see http://nces.ed.gov/ccd [November 2010].

TABLE 4-3 Rates of All ELL Students by State: School Years 2006-2007, 2007-2008, and 2008-2009

	2006-2007	2007-2008	2008-2009
Alabama	2.5	2.8	2.6
Alaska	15.7	12.8	9.2
Arizona	14.3	13.8	11.6
Arkansas	5.0	5.4	5.8
California	24.3	24.5	23.9
Colorado	11.3	10.6	11.1
Connecticut	4.6	5.3	5.2
Delaware	5.4	5.9	5.9
District of Columbia	6.5	6.5	7.5
Florida	8.8	8.7	8.5
Georgia	4.6	4.8	4.9
Hawaii	8.7	10.4	10.3
Idaho	6.2	6.1	6.5
Illinois	8.2	9.0	9.7
Indiana	4.1	4.4	4.4
Iowa	3.8	4.0	4.2
Kansas	6.2	6.8	7.3
Kentucky	1.6	1.9	2.2
Louisiana	1.3	1.7	1.8
Maine	1.9	2.1	2.1
Maryland	4.0	4.8	4.7
Massachusetts	5.6	5.8	5.1
Michigan	4.0	3.0	3.6
Minnesota	7.6	7.3	7.3
Mississippi	1.0	1.1	1.3
Missouri	2.4	2.1	1.8
Montana	4.8	4.7	3.2
Nebraska	6.3	6.6	6.3
Nevada	16.6	11.0	17.7
New Hampshire	1.5	1.6	2.0
New Jersey	3.9	3.9	3.9
New Mexico	18.5	18.6	16.4
New York	7.0	7.6	6.7
North Carolina	6.1	8.6	7.6
North Dakota	2.5	4.9	4.3
Ohio	1.6	1.9	2.0
Oklahoma	6.0	5.9	5.9
Oregon	11.0	11.0	11.1
Pennsylvania	2.4	2.6	2.6
Rhode Island	6.6	4.8	6.5
South Carolina	4.3	4.0	4.4
South Dakota	2.7	3.5	3.0
Tennessee	2.4	2.7	2.8
Texas	10.9	14.8	15.4
Utah	9.2	8.1	7.7
Vermont	1.8	1.6	1.6
Virginia	7.1	6.9	7.1
Washington	8.3	7.8	8.0
West Virginia	0.8	0.8	0.6
Wisconsin	4.7	5.0	5.5
Wyoming	3.5	2.8	2.6

NOTE: State rates of ELL students are calculated as the number of ELL students in the state divided by the number of K-12 students enrolled in public schools in the state.

SOURCE: Data from the U.S. Department of Education, Education Data Exchange Network except counts noted with an asterisk, which were obtained from the Consolidated State Performance Reports.

quite stable. However, there are still some fairly substantial changes. For example, Texas experienced an increase of about 5 percentage points in its rate of ELL students over the 3 school years, increasing from 10.9 percent in 2006-2007 to 15.4 percent in 2008-2009. In contract, over the same period, Alaska experienced a decrease of roughly 6 percentage points. The fluctuation in the absolute numbers of ELL students in Nevada also showed up in the data for the state rate, which was 16.6 percent in 2006-2007, 11.0 percent in 2007-2008, and 17.7 percent in 2008-2009.

Counts of Tested, Not Proficient Students

As noted above, states annually report the number of students who take their ELP test and the number of students who scored at the "English proficient" level.[10] Because these counts were not available through the EDEN system, staff at the DoEd provided us with the counts for two school years, 2007-2008 and 2008-2009:[11] see Table 4-4. The table lists the states in order by the average of the counts across the 2 school years. The table also shows each state's share of the total population of ELL students in the country who were tested and determined to be not proficient in English.

As with the counts of total ELL students (shown in Table 4-2 above), there were some fairly large differences in the absolute numbers across the 2 years. The largest increase was for Wisconsin, where the numbers more than tripled (from 12,865 students in 2007-2008 to 44,729 in 2008-2009). Colorado, Minnesota, and Virginia also saw fairly large increases, with their 2008-2009 counts roughly 50 percent higher than those for 2007-2008. In contrast, some states experienced decreases, the largest of which was Michigan, with a 31 percent decrease in absolute numbers from 2007-2008 to 2008-2009 (from 56,919 to 38,389).

Despite the fluctuations in absolute numbers, the states tended to rank order quite similarly across the 2 years with regard to their shares of students. Rank orderings of the seven states with the highest shares remained nearly identical across the 2 school years.

Comparison of state shares across the two types of counts (total ELL students and "tested, not proficient" students) also shows considerable similarity in the rank orderings. For instance, as can be seen by comparing Tables 4-2 and 4-4, the 16 states with the highest percentage shares are the same across the two counts; and the rank orderings of these states change only slightly across the two tables. For example, in Table 4-4, New York is ranked third and Florida is ranked fourth, and in Table 4-2,

[10] We determined the number of students who scored below the proficient level by finding the difference between these two numbers.

[11] These counts were not available for previous years because of differences in reporting and formatting. For the years of our analysis, counts were for all states except California for the 2008-2009 school year. In order to be able to include California in our analyses, we substituted the counts for the 2007-2008 school year for the missing data.

TABLE 4-4 Numbers and Shares of ELL Students Reported Tested, Not Proficient for 2007-2008 and 2008-2009 School Years

State	2007-2008 Number	2007-2008 Share[a]	2008-2009 Number	2008-2009 Share[a]	2-Year Average
California	896,521	29.37%	896,521[b]	28.63%	896,521.0
Texas	460,680	15.09%	477,611	15.25%	469,145.5
New York	170,710	5.59%	166,212	5.31%	168,461.0
Florida	165,325	5.42%	161,734	5.16%	163,529.5
Arizona	126,675	4.15%	89,555	2.86%	108,115.0
Illinois	108,836	3.56%	101,507	3.24%	105,171.5
North Carolina	101,645	3.33%	93,150	2.97%	97,397.5
Colorado	53,340	1.75%	84,660	2.70%	69,000.0
Washington	67,425	2.21%	69,631	2.22%	68,528.0
Virginia	52,910	1.73%	83,538	2.67%	68,224.0
Nevada	63,642	2.08%	66,330	2.12%	64,986.0
Georgia	62,576	2.05%	62,999	2.01%	62,787.5
Oregon	55,390	1.81%	55,301	1.77%	55,345.5
Michigan	56,919	1.86%	38,389	1.23%	47,654.0
New Mexico	44,874	1.47%	43,824	1.40%	44,349.0
Minnesota	35,871	1.17%	52,452	1.67%	44,161.5
Indiana	38,334	1.26%	41,569	1.33%	39,951.5
New Jersey	38,953	1.28%	40,571	1.30%	39,762.0
Pennsylvania	36,007	1.18%	31,886	1.02%	33,946.5
Massachusetts	26,212	0.86%	36,354	1.16%	31,283.0
Ohio	29,584	0.97%	31,267	1.00%	30,425.5
Oklahoma	29,484	0.97%	28,477	0.91%	28,980.5
Wisconsin	12,865	0.42%	44,729	1.43%	28,797.0
Kansas	28,455	0.93%	27,003	0.86%	27,729.0
Utah	27,733	0.91%	27,666	0.88%	27,699.5
South Carolina	26,147	0.86%	27,937	0.89%	27,042.0
Maryland	19,718	0.65%	33,518	1.07%	26,618.0
Arkansas	23,612	0.77%	25,104	0.80%	24,358.0
Tennessee	19,376	0.63%	18,588[b]	0.59%	18,982.0
Connecticut	18,535	0.61%	16,881	0.54%	17,708.0
Hawaii	15,085	0.49%	15,649	0.50%	15,367.0
Alabama	16,099	0.53%	12,490	0.40%	14,294.5
Missouri	12,185	0.40%	16,313	0.52%	14,249.0
Iowa	14,203	0.47%	14,197	0.45%	14,200.0
Alaska	14,183	0.46%	13,861	0.44%	14,022.0
Idaho	14,157	0.46%	10,530	0.34%	12,343.5
Nebraska	12,244	0.40%	12,044	0.38%	12,144.0
Kentucky	11,493	0.38%	12,771	0.41%	12,132.0
Louisiana	11,456	0.38%	10,206	0.33%	10,831.0
Rhode Island	5,741	0.19%	6,505	0.21%	6,123.0
District of Columbia	4,656	0.15%	4,664	0.15%	4,660.0
Delaware	3,089	0.10%	4,999	0.16%	4,044.0

continued

TABLE 4-4 Continued

State	2007-2008 Number	2007-2008 Share[a]	2008-2009 Number	2008-2009 Share[a]	2-Year Average
Maine	3,134	0.10%	3,789	0.12%	3,461.5
Mississippi	1,159	0.04%	5,645	0.18%	3,402.0
New Hampshire	2,840	0.09%	3,348	0.11%	3,094.0
North Dakota	4,257	0.14%	1,923	0.06%	3,090.0
South Dakota	2,846	0.09%	2,818	0.09%	2,832.0
Wyoming	1,872	0.06%	1,856	0.06%	1,864.0
Montana	1,572	0.05%	1,052	0.03%	1,312.0
Vermont	1,210	0.04%	1,208	0.04%	1,209.0
West Virginia	1,148	0.04%	853	0.03%	1,000.5
United States	43,052,983		3,131,685		

NOTES: The numbers of tested, not proficient students were computed for each state by subtracting the number of all LEP (ELL) students proficient or above on a state's proficiency test from the number of all LEP (ELL) students tested on the state annual ELP assessment. States are listed in order by the 2-year average of their numbers of ELL students determined to be tested, not proficient.

[a]Percentages represent each state's share of the tested, not proficient students in the country.

[b]Data not available; 2007-2008 count used so that state shares could be estimated.

SOURCE: Data from the Consolidated State Performance Reports (CSPR) provided by the U.S. Department of Education.

Florida is third and New York is fourth. Similarly, in Table 4-4, Arizona is fifth and Illinois is sixth, and in Table 4-2, Illinois is fifth and Arizona is sixth.

These small differences in shares indicate that allocations based on the counts of all ELL students and allocations based on the counts of tested, not proficient students would be quite similar. The overall correlations between the shares for the two counts were 0.99 for both 2007-2008 and 2008-2009.

Table 4-5 shows the state rates for the counts of tested, not proficient students. The rates in Table 4-5 show the count of tested, not proficient students as a percentage of the total population of school-age children in the state. For the most part, the state rates are similar across the 2 school years, with fluctuations generally in the range of 2-3 percentage points. The largest difference was in Wisconsin, where the increase in absolute numbers across the 2 school years resulted in an increase of the rate from 1.47 percent to 5.12 percent. Comparison of state rates across the two types of counts (Tables 4-3 and 4-5) reveals some differences: the overall correlations between the rates for the two counts were 0.92 for 2007-2008 and 0.95 for 2008-2009.

Effect of Data Reporting Systems on Data Quality

Although we have documented some anomalies in the state-provided counts of ELL students, they seem to be less prevalent in data for the most recent school

TABLE 4-5 Rates of Tested, Not Proficient Students by State, 2007-2008 and 2008-2009 School Years (in percentage*)

State	2007-2008	2008-2009
Alabama	2.16	1.68
Alaska	10.82	10.61
Arizona	11.65	8.23
Arkansas	4.93	5.24
California	14.13	N/A
Colorado	6.65	10.34
Connecticut	3.25	2.98
Delaware	2.52	3.99
District of Columbia	5.94	6.79
Florida	6.20	6.15
Georgia	3.79	3.80
Hawaii	8.39	8.72
Idaho	5.20	3.83
Illinois	5.15	4.79
Indiana	3.66	3.97
Iowa	2.93	2.91
Kansas	6.08	5.73
Kentucky	1.73	1.91
Louisiana	1.68	1.49
Maine	1.60	1.97
Maryland	2.33	3.97
Massachusetts	2.72	3.79
Michigan	3.36	2.31
Minnesota	4.28	6.27
Mississippi	0.23	1.15
Missouri	1.33	1.78
Montana	1.10	0.74
Nebraska	4.20	4.12
Nevada	14.82	15.31
New Hampshire	1.41	1.69
New Jersey	2.82	2.94
New Mexico	13.64	13.27
New York	6.17	6.06
North Carolina	6.82	6.26
North Dakota	4.48	2.03
Ohio	1.62	1.72
Oklahoma	4.59	4.41
Oregon	9.79	9.82
Pennsylvania	2.00	1.80
Rhode Island	3.89	4.48
South Carolina	3.67	3.89
South Dakota	2.34	2.23
Tennessee	2.01	1.91

continued

TABLE 4-5 Continued

State	2007-2008	2008-2009
Texas	9.85	10.05
Utah	4.81	4.94
Vermont	1.29	1.31
Virginia	4.30	6.76
Washington	6.54	6.71
West Virginia	0.41	0.30
Wisconsin	1.47	5.12
Wyoming	2.17	2.13

*Percentages are calculated as the number of test not proficient students in the state divided by the number of K-12 students enrolled in public schools in the state.
SOURCES: Data are from the Consolidated State Performance Reports (CSPR) provided by the U.S. Department of Education. The numbers of tested, not proficient students were computed for each state by subtracting the number of all LEP students proficient or above on the state annual ELP assessment from the number of all LEP students tested on the state annual ELP assessment.

year than in prior years. For instance, counts of total ELL students were missing for five states for the 2006-2007 school year but for only one state for the 2008-2009 school year. And for the 2 years we examined, only one state had missing data for the count of tested, not proficient students. The DoEd staff told us that they have worked on refining the instructions and formatting of the CSPRs and on ensuring that the EDEN data are accurate. The department has worked to remediate some of the earlier problems and developed a format for data elements that has been consistent for several years now.

We also note that, in education, as in many other fields, there have been vast improvements in data availability and access in recent years. The local and state education agencies and the DoEd have been particularly driven to improve data to address new accountability provisions of the NCLB.

Other factors have also influenced the recent advances in data availability, access, and quality. Standardization has been enhanced over the years by efforts such as the National Center for Education Statistics initiative to support the National Forum on Education Statistics, which brings together data system specialists from state agencies to focus on common data issues. The attention that has been paid to documenting the EDEN system and clarifying its specifications has also paid dividends. New initiatives—such as efforts to create quality longitudinal databases represented in the work of the Data Quality Campaign and recently augmented by $250 million in American Recovery and Reinvestment Act funding—have also begun to pay dividends in standardization of data elements and the development of sophisticated data systems to capture, analyze, and promulgate student data.

Thus, while we find some issues with the quality and availability of data on ELL students, we recognize the significant improvements that have been made throughout the system, and are encouraged by new initiatives that continue to focus attention on further improvements.

CONCLUSION 4-4 There are concerns about the accuracy of the compilation and reporting of state data to the Department of Education. However, there have been significant improvements in the collection and reporting of these data over the past several years, and systems show promise for further improvements in the coming years.

5

Comparison of American Community Survey Estimates and State Counts

The previous chapters have described various aspects of the American Community Survey (ACS) and have documented state policies, practices, and criteria that affect the counts of English language learner (ELL) students that are reported by state education agencies. It is readily apparent that these two allowable sources of data for use in allocating Title III funds to states have distinct strengths and weaknesses. In this chapter, we first briefly discuss the concepts and methods that underlie the two counts. We then consider in depth the differences between and the ACS estimates and state-provided counts on several dimensions.

CONCEPTUAL DIFFERENCES IN THE TWO SOURCES

The ACS estimates and state-provided counts of ELL students are two very different mechanisms for determining the number of school-age children in a state likely to have difficulty with English. As shown in Table 5-1, they differ along a number of dimensions.

The ACS is an indirect and subjective measure in that a parent or other adult household member provides an assessment for each child in the home. Since the question only asks about spoken English, it focuses on a single modality, and no context for English use is specified. The respondent may be considering the child's proficiency with English in any number of settings (i.e., family life, community, social, academic), and the child may have different levels of proficiency in different settings. However, the ACS questions and criteria are consistent across states.

In contrast, the state-provided counts are based on direct, relatively objective measures of students' English language proficiency. The counts are based on comprehensive processes established by state and local education agencies that consider

TABLE 5-1 Differences Between the ACS Estimates and State-Provided Counts of ELL Students

Type of Difference	ACS Estimate	State-Provided Count
Age Range	5-21 years of age	Not specified (elementary and secondary school-aged population is usually defined as 5-18 years of age)
School Enrollment	Enrollment status not specified (i.e., includes public and private schools)	Newly and continually enrolled in elementary and secondary schools for which Consolidated State Performance Reports are submitted by state education agencies (i.e., public schools including charter schools)
Assessment Method	Single question regarding spoken English ability	Comprehensive assessment that incorporates information from multiple sources
Mode of Response	Indirect and subjective measure, based on the response of a parent (or other adult in the household) to a single question	Direct evaluation based on a student's performance in acquiring English proficiency
Modality(ies) Assessed	Speaking	Speaking, listening, reading, and writing
Context Assessed	Not specified: likely to be community and family setting	Classroom setting
Basis for Distinguishing Proficient from Not Proficient	Single national cut score	State- or local-determined criteria
Comparability Across States	Item is identically presented across the nation; estimates based on a uniform methodology across the states	States use different assessments, procedures, cut scores, and criteria; estimates based on different methodologies

language proficiency across multiple modalities (listening, speaking, reading, and writing). The measures of language proficiency explicitly address both academic and social contexts. Unlike the ACS estimates, the information from the states varies because the policies, practices, and criteria used by the states are not uniform.

CONCLUSION 5-1 The criteria used by the states for counts of English language learner students are more conceptually sound than the criteria on which American Community Survey (ACS) estimates are based. However, the policies, practices, and criteria used by the states differ from state to

state, while the ACS provides estimates on the basis of a uniform methodology across the country.

Despite their differences, the ACS estimates and state counts represent conceptually similar entities—both measure the number of school-age children in the state that have not mastered English. Thus, some level of correspondence between the two measures would be expected.

We conducted a series of analyses to evaluate the consistency of the ACS and state-provided percentages of ELL students. In order to facilitate comparisons between the ACS estimates and state-provided counts we limit the ACS population to those aged 5-18 and only to those enrolled in public school. It is important to point out that by limiting the ACS estimate to this comparison group, we have created an ACS-based variable that is more limited than the legal definition of ELL students used by the U.S. Department of Education (DoEd).

As detailed in Chapters 2 and 4, there are two ways to calculate the percentages: the number of ELL children in the state as a percentage of the total number of ELL children in the country, which is the state's *share* of ELL children; and the proportion that ELL students constitute of the total number of enrolled students, which is the *rate* of ELL students. We also conducted a series of multiple regression analyses to evaluate the correspondence between the ACS and state estimates. In these analyses, we focus on rates, rather than shares, in order to assess the degree of consistency of the two data sources in a manner that is relatively independent of state population. That is, analyses that focus on state counts or shares are dominated by the agreement between the ACS and state-provided numbers for some states, suggesting only that certain states (notably, California and Texas) are large and others are small, a trivial finding that provides little information about how well the two measures agree on estimation of ELL students.

COMPARISON OF SHARES OF ELL STUDENTS

In this section we compare the state shares (of Title III funding) based on ACS estimates with those based on state-provided counts. Since the funding allocations are based on each state's share of ELL students in the country, this analysis allows us to evaluate how the allocations would be affected on the basis of which measure was used, as well as the ways that the measures would result in different funding decisions. We compare the shares in three ways: (1) the percentage shares themselves, (2) the ratio of the shares, and (3) the absolute differences in the shares across the states.

State Percentage Shares

Table 5-2 shows each state's share of ELL students based on the two data sources. The first four columns on the left-hand side of the table show the shares based on the ACS estimates. Included are 1-year estimates for 2006, 2007, and 2008

TABLE 5-2 Shares of ELL Students Based on ACS and State-Provided Counts (in percentage)

	ACS Estimate				State-Provided Count				
					All ELL[a]			Tested Not, Proficient[b]	
State	2006	2007	2008	2006-2008	2006-2007	2007-2008	2008-2009	2007-2008	2008-2009
Alabama	0.43	0.40	0.40	0.40	0.43	0.46	0.43	0.53	0.40
Alaska	0.17	0.17	0.15	0.17	0.48	0.37	0.27	0.46	0.44
Arizona	3.66	3.90	3.69	3.74	3.56	3.31	2.79	4.15	2.86
Arkansas	0.45	0.48	0.37	0.44	0.55	0.57	0.61	0.77	0.80
California[c]	29.12	28.31	27.25	28.12	36.35	34.32	33.68	29.37	28.63[c]
Colorado	1.70	1.74	1.66	1.70	2.10	1.89	1.98	1.75	2.70
Connecticut	0.88	0.65	0.58	0.71	0.61	0.66	0.66	0.61	0.54
Delaware	0.17	0.16	0.15	0.16	0.15	0.16	0.16	0.10	0.16
District of Columbia	0.08	0.06	0.07	0.06	0.11	0.11	0.13	0.15	0.15
Florida	5.54	5.51	5.25	5.38	5.47	5.11	5.03	5.42	5.16
Georgia	2.10	2.20	1.97	2.09	1.73	1.77	1.80	2.05	2.01
Hawaii	0.30	0.22	0.38	0.29	0.37	0.41	0.41	0.49	0.50
Idaho	0.28	0.25	0.32	0.29	0.39	0.37	0.39	0.46	0.34
Illinois	4.52	4.81	4.74	4.67	4.03	4.20	4.55	3.56	3.24
Indiana	0.91	0.81	0.89	0.88	0.99	1.02	1.02	1.26	1.33
Iowa	0.44	0.40	0.40	0.43	0.42	0.43	0.45	0.47	0.45
Kansas	0.51	0.49	0.54	0.52	0.67	0.70	0.76	0.93	0.86
Kentucky	0.31	0.39	0.45	0.41	0.25	0.28	0.32	0.38	0.41
Louisiana	0.26	0.38	0.39	0.36	0.20	0.25	0.28	0.38	0.33
Maine	0.14	0.11	0.08	0.12	0.09	0.09	0.09	0.10	0.12
Maryland	0.96	1.10	0.98	1.03	0.80	0.89	0.89	0.65	1.07
Massachusetts	1.71	1.57	1.61	1.64	1.26	1.23	1.09	0.86	1.16
Michigan	1.56	1.51	1.40	1.49	1.63	1.14	1.35	1.86	1.23
Minnesota	1.18	1.22	1.35	1.27	1.49	1.35	1.37	1.17	1.67
Mississippi	0.18	0.16	0.16	0.20	0.12	0.12	0.15	0.04	0.18

Missouri	0.56	0.69	0.60	0.64	0.52	0.42	0.36	0.40	0.52
Montana	0.06	0.05	0.04	0.06	0.16	0.15	0.10	0.05	0.03
Nebraska	0.41	0.37	0.39	0.39	0.42	0.42	0.41	0.40	0.38
Nevada	1.02	1.15	1.38	1.19	1.64	1.04	1.69	2.08	2.12
New Hampshire	0.06	0.11	0.09	0.09	0.07	0.07	0.09	0.09	0.11
New Jersey	2.55	2.29	2.57	2.48	1.27	1.20	1.20	1.28	1.30
New Mexico	0.97	0.84	0.74	0.83	1.42	1.35	1.20	1.47	1.40
New York	6.63	6.36	6.80	6.61	4.57	4.65	4.09	5.59	5.31
North Carolina	2.06	1.88	2.25	2.04	2.05	2.82	2.53	3.33	2.97
North Dakota	0.04	0.07	0.06	0.05	0.06	0.10	0.09	0.14	0.06
Ohio	1.03	0.93	1.02	1.00	0.68	0.77	0.81	0.97	1.00
Oklahoma	0.45	0.53	0.47	0.50	0.89	0.83	0.85	0.97	0.91
Oregon	1.25	1.23	1.14	1.21	1.44	1.37	1.40	1.81	1.77
Pennsylvania	1.48	1.52	1.60	1.53	1.06	1.02	1.06	1.18	1.02
Rhode Island	0.19	0.29	0.27	0.25	0.23	0.16	0.16	0.19	0.21
South Carolina	0.51	0.57	0.51	0.56	0.70	0.63	0.70	0.86	0.89
South Dakota	0.05	0.05	0.09	0.07	0.08	0.09	0.08	0.09	0.09
Tennessee	0.61	0.67	0.66	0.68	0.54	0.57	0.61	0.63	0.59
Texas	16.98	17.75	18.55	17.63	11.69	15.31	15.97	15.09	15.25
Utah	0.75	0.77	0.71	0.75	1.13	1.03	0.99	0.91	0.88
Vermont	0.04	0.03	0.03	0.03	0.04	0.03	0.03	0.04	0.04
Virginia	1.57	1.22	1.45	1.42	2.01	1.86	1.93	1.73	2.67
Washington	2.04	2.31	2.35	2.22	1.98	1.78	1.84	2.21	2.22
West Virginia	0.12	0.09	0.09	0.11	0.05	0.05	0.04	0.04	0.03
Wisconsin	0.96	1.18	0.86	1.01	0.96	0.97	1.06	0.42	1.43
Wyoming	0.04	0.03	0.05	0.04	0.07	0.05	0.05	0.06	0.06
Total Count in the U.S.	2,491,160	2,482,420	2,433,845	2,462,330	4,289,046	4,525,892	4,499,072	3,052,983	3,131,685

[a]The total number of ELL students was obtained from the EDEN database.

[b]The number of tested, not proficient students was computed for each state from the state Consolidated State Performance Reports by subtracting the number of all LEP students proficient or above on state annual ELP assessments (1.6.3.1.2) from the number of all LEP students tested on state annual ELP assessments (1.6.3.1.1).

[c]Counts for California for 2008-2009 were unavailable, percentage is based on the 2007-2008 count.

and the 3-year estimate across these years. The table includes two types of shares calculated from the state-provided counts. Three of the columns show the state-provided counts of all ELL students for the 2006-2007, 2007-2008, and 2008-2009 school years. The other two columns show the shares based on the state-provided counts of ELL students who were determined to be not proficient on the English language proficiency (ELP) test for the 2007-2008 and 2008-2009 school years.

Ratios of the Shares

To help compare the percentages from the two measures, we calculated the ratio of the share based on the ACS estimate to the share based on the state-provided counts. These ratios are shown in Table 5-3. The tables include the ratios of the ACS 1-year estimate to the state counts for each school year, as well as the ratios of the ACS 3-year estimate to the most recent state school ycar data.

Ratios higher than 1.00 indicate that the share based on the ACS estimate was higher than the share based on the state-provided count, and ratios that are less than 1.00 indicate that the state-provided count was higher than the ACS estimate. Scanning the ratios across the time spans and the type of state-provided counts reveals considerable consistency. That is, for a given state, the ratios were generally consistently above 1.00 (ACS estimate higher than state-provided count) or consistently below 1.00 (state-provided count higher than ACS estimate). It is difficult to discern any explanatory factors from this comparison. No patterns appear to be evident due to region of the country or type of proficiency test used. For example, about half of the states that used the ACCESS for ELLs test developed by the World-Class Instructional Design and Assessment Consortium (see Table 4-1 in Chapter 4) had ACS rates higher than the state rates and half were lower.

Absolute Differences in the Shares

To quantify the potential effects of the differences between the two data sources in terms of the distribution of Title III funds, we calculated the total absolute differences between the shares based on ACS estimates and those based on state-provided counts. The differences are shown in Table 5-4. The left-hand side of the table shows the values for the differences between ACS estimates and the state-provided counts of all ELL students, for both 1-year and 3-year ACS estimates; the right-hand side of the table shows the values of the differences between ACS estimates and the state-provided counts of tested, not proficient students.

This quantity varies from about 20 percent to 26 percent of the total allocation, depending on the years considered. Because every dollar moved is counted twice in this total (once when it is taken from a state with a reduced share and once when added to one with an increased share), it means that from 10 percent to 13 percent of the total dollars would be moved by switching from one allocation to another, a substantial change in allocations.

TABLE 5-3 Ratio of State Shares Based on ACS Estimate to Shares Based on State-Provided Counts

	All ELL Students[a]				Tested, Not Proficient Students[b]		
State	ACS 2006 to State 2006-2007	ACS 2007 to State 2007-2008	ACS 2008 to State 2008-2009	ACS 2006-2008 to State 2008-2009	ACS 2007 to State 2007-2008	ACS 2008 to State 2008-2009	ACS 2006-2008 to State 2008-2009
Alabama	1.00	0.87	0.91	0.91	0.76	0.99	0.99
Alaska	0.35	0.47	0.56	0.65	0.37	0.34	0.39
Arizona	1.03	1.18	1.32	1.34	0.94	1.29	1.31
Arkansas	0.81	0.84	0.61	0.72	0.62	0.47	0.55
California[c]	0.80	0.82	0.81	0.83	0.96	0.95	0.98[c]
Colorado	0.81	0.93	0.84	0.86	1.00	0.61	0.63
Connecticut	1.43	0.98	0.87	1.08	1.07	1.07	1.33
Delaware	1.10	1.01	0.93	1.02	1.60	0.93	1.02
District of Columbia	0.71	0.56	0.51	0.46	0.41	0.45	0.40
Florida	1.01	1.08	1.05	1.07	1.02	1.02	1.04
Georgia	1.21	1.25	1.10	1.16	1.07	0.98	1.04
Hawaii	0.83	0.54	0.92	0.71	0.45	0.76	0.58
Idaho	0.72	0.68	0.80	0.74	0.54	0.94	0.87
Illinois	1.12	1.15	1.04	1.03	1.35	1.46	1.44
Indiana	0.92	0.79	0.88	0.87	0.65	0.67	0.66
Iowa	1.04	0.93	0.89	0.96	0.86	0.89	0.96
Kansas	0.76	0.69	0.71	0.69	0.52	0.62	0.61
Kentucky	1.22	1.38	1.40	1.27	1.04	1.11	1.01
Louisiana	1.27	1.52	1.41	1.30	1.02	1.20	1.11
Maine	1.68	1.19	0.88	1.30	1.03	0.68	1.01
Maryland	1.20	1.23	1.10	1.16	1.70	0.92	0.96
Massachusetts	1.36	1.28	1.48	1.50	1.83	1.39	1.41
Michigan	0.96	1.32	1.04	1.10	0.81	1.15	1.22
Minnesota	0.79	0.90	0.99	0.93	1.04	0.81	0.76
Mississippi	1.53	1.34	1.07	1.37	4.24	0.86	1.11
Missouri	1.08	1.63	1.66	1.77	1.72	1.16	1.23
Montana	0.35	0.34	0.39	0.61	0.99	1.19	1.84

continued

TABLE 5-3 Continued

State	All ELL Students[a]				Tested, Not Proficient Students[b]		
	ACS 2006 to State 2006-2007	ACS 2007 to State 2007-2008	ACS 2008 to State 2008-2009	ACS 2006-2008 to State 2008-2009	ACS 2007 to State 2007-2008	ACS 2008 to State 2008-2009	ACS 2006-2008 to State 2008-2009
Nebraska	0.96	0.87	0.96	0.96	0.92	1.02	1.02
Nevada	0.62	1.10	0.82	0.71	0.55	0.65	0.56
New Hampshire	0.78	1.55	0.98	0.99	1.18	0.83	0.84
New Jersey	2.01	1.90	2.13	2.06	1.79	1.98	1.91
New Mexico	0.68	0.62	0.62	0.69	0.57	0.53	0.59
New York	1.45	1.37	1.66	1.62	1.14	1.28	1.25
North Carolina	1.01	0.67	0.89	0.81	0.56	0.76	0.69
North Dakota	0.69	0.66	0.63	0.56	0.49	0.92	0.82
Ohio	1.51	1.20	1.27	1.23	0.96	1.03	1.00
Oklahoma	0.51	0.64	0.56	0.59	0.55	0.52	0.55
Oregon	0.86	0.90	0.82	0.87	0.68	0.65	0.69
Pennsylvania	1.39	1.48	1.51	1.45	1.29	1.58	1.50
Rhode Island	0.82	1.85	1.71	1.59	1.55	1.31	1.22
South Carolina	0.73	0.91	0.73	0.81	0.66	0.57	0.63
South Dakota	0.71	0.56	1.12	0.91	0.56	1.00	0.81
Tennessee	1.14	1.19	1.09	1.11	1.06	1.12	1.14
Texas	1.45	1.16	1.16	1.10	1.18	1.22	1.16
Utah	0.66	0.75	0.72	0.76	0.85	0.81	0.85
Vermont	0.96	0.88	0.94	0.98	0.72	0.81	0.84
Virginia	0.78	0.65	0.75	0.73	0.70	0.54	0.53
Washington	1.03	1.30	1.28	1.21	1.05	1.06	1.00
West Virginia	2.35	1.83	2.46	3.03	2.51	3.24	4.01
Wisconsin	1.00	1.21	0.81	0.95	2.79	0.60	0.71
Wyoming	0.56	0.54	0.95	0.83	0.47	0.81	0.71

[a]The total number of ELL students was obtained from the Education Data Exchange Network (EDEN) database.

[b]The number of tested, not proficient students was computed for each state from the state Consolidated State Performance Reports by subtracting the number of all LEP (ELL) students proficient or above on state annual ELP assessments (1.6.3.1.2) from the number of all LEP (ELL) students tested on state annual ELP assessments (1.6.3.1.1).

[c]Counts for California for 2008-2009 were unavailable; the percentage is based on the 2007-2008 count.

TABLE 5-4 Total Absolute Difference Between Shares Based on ACS Estimates and Shares Based on State-Provided Counts

	ALL ELL Students			Tested, Not Proficient Students		
Type of ACS Estimate	ACS 2006 and State 2006-2007	ACS 2007 and State 2007-2008	ACS 2008 and State 2008-2009	ACS 2006-2008 and State 2006-2007	ACS 2006-08 and State 2007-2008	ACS 2006-2008 and State 2008-2009
1-year	23.68	21.13	21.57	N/A	18.21	20.55
3-year	25.94	20.44	19.73	N/A	17.94	18.66

COMPARISON OF RATES OF ELL STUDENTS

As noted above, comparison of state rates removes the simple effect of size from the analyses and thereby focuses attention on differences in measurement.

State Rates

Table 5-5 shows each state's rate of ELL students based on the two data sources. The first four columns on the left-hand side of the table show the rates based on the ACS estimates, including 1-year estimates for 2006, 2007, and 2008 and the 3-year estimate across these years. For the state-provided counts, two types of rates calculated are shown: the rates based on state-provided counts of all ELL students for the 2006-2007, 2007-2008, and 2008-2009 school years and the rates based on the state-provided counts of tested, not proficient students for the 2007-2008 and 2008-2009 school years.

The rates derived from the ACS were lower than the rates derived from state-provided counts in all but two states (New Jersey and West Virginia), when state-provided counts were based on all ELL students. When the state-provided count was based on the number of tested, not proficient students, the ACS estimates were consistently lower in five states (Illinois, Massachusetts, New Jersey, Pennsylvania, and West Virginia). In the most recent period, the average percentage of ELL students for the nation was about 5 percent for the ACS and about 9 percent for the state-provided rates; for the state-provided rate of tested, not proficient students, the rate was 6 percent.

Ratio of the Rates

To compare the percentages from the two measures, we calculated the ratio of the rate based on the ACS estimate to the rate based on each of the state-provided counts. These ratios are shown in Table 5-6. The left-hand side of the table shows

TABLE 5-5 Rate of ELL Students by State Based on ACS Estimates and State-Provided Counts (in percentage)

	ACS Estimate				State-Provided Count				
					All ELL[a]			Tested, Not Proficient[b]	
State	2006	2007	2008	2006-2008	2006-2007	2007-2008	2008-2009	2007-2008	2008-2009
Alabama	1.44	1.36	1.31	1.33	2.47	2.81	2.62	2.16	1.68
Alaska	3.34	3.56	3.09	3.53	15.66	12.84	9.21	10.82	10.61
Arizona	8.50	8.82	8.01	8.40	14.30	13.77	11.55	11.65	8.23
Arkansas	2.40	2.54	1.94	2.32	4.96	5.41	5.77	4.93	5.24
California[c]	11.34	11.13	10.54	11.00	24.34	24.48	24.23	14.13	14.34[c]
Colorado	5.55	5.53	5.14	5.37	11.32	10.64	10.86	6.65	10.34
Connecticut	3.90	2.91	2.55	3.16	4.58	5.26	5.25	3.25	2.98
Delaware	3.51	3.20	2.93	3.26	5.44	5.92	5.73	2.52	3.99
District of Columbia	2.91	2.60	2.57	2.34	6.47	6.54	8.52	5.94	6.79
Florida	5.35	5.33	4.99	5.16	8.78	8.68	8.59	6.20	6.15
Georgia	3.24	3.31	2.89	3.15	4.55	4.85	4.89	3.79	3.80
Hawaii	4.40	3.40	5.60	4.36	8.66	10.38	10.34	8.39	8.72
Idaho	2.69	2.30	2.84	2.70	6.25	6.13	6.42	5.20	3.83
Illinois	5.41	5.72	5.57	5.54	8.16	8.99	9.66	5.15	4.79
Indiana	2.24	1.98	2.12	2.13	4.07	4.42	4.37	3.66	3.97
Iowa	2.24	2.07	2.03	2.22	3.75	4.01	4.17	2.93	2.91
Kansas	2.75	2.60	2.85	2.81	6.16	6.78	7.24	6.08	5.73
Kentucky	1.18	1.48	1.67	1.54	1.58	1.94	2.18	1.73	1.91
Louisiana	0.96	1.44	1.40	1.29	1.28	1.68	1.82	1.68	1.49
Maine	1.81	1.35	1.05	1.55	1.90	2.06	2.19	1.60	1.97
Maryland	2.85	3.25	2.87	3.02	4.03	4.78	4.75	2.33	3.97
Massachusetts	4.51	4.16	4.15	4.27	5.58	5.79	5.12	2.72	3.79
Michigan	2.27	2.23	2.06	2.18	4.05	3.04	3.67	3.36	2.31
Minnesota	3.52	3.67	4.07	3.81	7.60	7.31	7.35	4.28	6.27
Mississippi	0.87	0.79	0.74	0.98	1.01	1.10	1.33	0.23	1.15

Missouri	1.57	1.88	1.63	1.75	2.43	2.08	1.78	1.33	1.78
Montana	0.94	0.86	0.66	1.02	4.84	4.71	3.21	1.10	0.74
Nebraska	3.56	3.22	3.40	3.42	6.33	6.57	6.29	4.20	4.12
Nevada	5.99	6.45	7.62	6.79	16.61	10.96	17.53	14.82	15.31
New Hampshire	0.70	1.35	1.09	1.09	1.55	1.59	2.06	1.41	1.69
New Jersey	4.68	4.26	4.65	4.56	3.92	3.94	3.92	2.82	2.94
New Mexico	7.03	6.26	5.43	6.14	18.50	18.60	16.34	13.64	13.27
New York	5.72	5.59	5.87	5.72	6.98	7.61	6.71	6.17	6.06
North Carolina	3.54	3.19	3.73	3.47	6.07	8.56	7.65	6.82	6.26
North Dakota	0.95	1.70	1.44	1.27	2.48	4.89	4.29	4.48	2.03
Ohio	1.43	1.29	1.42	1.38	1.59	1.92	2.00	1.62	1.72
Oklahoma	1.84	2.17	1.89	2.02	5.96	5.88	5.90	4.59	4.41
Oregon	5.47	5.40	4.90	5.28	11.01	10.98	11.19	9.79	9.82
Pennsylvania	2.05	2.11	2.22	2.13	2.43	2.57	2.69	2.00	1.80
Rhode Island	3.11	4.71	4.52	4.12	6.62	4.84	6.47	3.89	4.48
South Carolina	1.84	2.03	1.77	1.99	4.26	3.98	4.38	3.67	3.89
South Dakota	1.06	0.99	1.66	1.38	2.72	3.47	2.84	2.34	2.23
Tennessee	1.62	1.76	1.71	1.76	2.35	2.66	2.82	2.01	1.91
Texas	9.69	10.02	10.07	9.89	10.90	14.82	15.12	9.85	10.05
Utah	3.54	3.52	3.15	3.42	9.25	8.12	7.94	4.81	4.94
Vermont	1.03	0.77	0.85	0.87	1.83	1.55	1.62	1.29	1.31
Virginia	3.28	2.51	2.94	2.92	7.08	6.85	7.04	4.30	6.76
Washington	4.94	5.64	5.60	5.38	8.26	7.83	7.98	6.54	6.71
West Virginia	1.13	0.87	0.79	0.99	0.80	0.83	0.57	0.41	0.30
Wisconsin	2.82	3.39	2.45	2.91	4.71	5.01	5.48	1.47	5.12
Wyoming	1.17	0.82	1.34	1.20	3.53	2.77	2.61	2.17	2.13
United States	5.10	5.10	5.00	5.10	8.70	9.20	9.10	6.19	6.37

[a]The total number of ELL students was obtained from the EDEN database.

[b]The total number of tested, not proficient was computed for each state from the state Consolidated State Performance Reports by subtracting the number of all LEP students proficient or above on state annual ELP assessments (1.6.3.1.2) from the number of all LEP students tested on state annual ELP assessments (1.6.3.1.1).

[c]Counts for California for 2008-2009 were unavailable; percentage is based on the 2007-2008 count.

TABLE 5-6 Ratio of Rates Based on ACS Estimates to Rates Based on State-Provided Counts

	All ELL Students[a]				Tested, Not Proficient Students[b]		
State	ACS 2006 to State 2006-2007	ACS 2007 to State 2007-2008	ACS 2008 to State 2008-2009	ACS 2006-2008 to State 2008-2009	ACS 2007 to State 2007-2008	ACS 2008 to State 2008-2009	ACS 2006-2008 to State 2008-2009
Alabama	0.58	0.48	0.50	0.51	0.63	0.78	0.80
Alaska	0.21	0.28	0.34	0.38	0.33	0.29	0.33
Arizona	0.59	0.64	0.69	0.73	0.76	0.97	1.02
Arkansas	0.48	0.47	0.34	0.40	0.52	0.37	0.44
California[c]	0.47	0.45	0.44	0.45	0.79	0.74[c]	0.77[c]
Colorado	0.49	0.52	0.47	0.49	0.83	0.50	0.52
Connecticut	0.85	0.55	0.49	0.60	0.89	0.86	1.06
Delaware	0.64	0.54	0.51	0.57	1.27	0.74	0.82
District of Columbia	0.45	0.40	0.30	0.27	0.44	0.38	0.34
Florida	0.61	0.61	0.58	0.60	0.86	0.81	0.84
Georgia	0.71	0.68	0.59	0.64	0.87	0.76	0.83
Hawaii	0.51	0.33	0.54	0.42	0.41	0.64	0.50
Idaho	0.43	0.37	0.44	0.42	0.44	0.74	0.70
Illinois	0.66	0.64	0.58	0.57	1.11	1.16	1.16
Indiana	0.55	0.45	0.49	0.49	0.54	0.53	0.54
Iowa	0.60	0.52	0.49	0.53	0.71	0.70	0.76
Kansas	0.45	0.38	0.39	0.39	0.43	0.50	0.49
Kentucky	0.75	0.76	0.77	0.71	0.86	0.88	0.81
Louisiana	0.75	0.85	0.77	0.71	0.86	0.94	0.87
Maine	0.95	0.66	0.48	0.71	0.85	0.54	0.79
Maryland	0.71	0.68	0.60	0.64	1.40	0.72	0.76
Massachusetts	0.81	0.72	0.81	0.83	1.53	1.09	1.13
Michigan	0.56	0.73	0.56	0.59	0.66	0.89	0.94
Minnesota	0.46	0.50	0.55	0.52	0.86	0.65	0.61
Mississippi	0.86	0.71	0.56	0.73	3.35	0.65	0.85

Missouri	0.64	0.90	0.92	0.99	1.41	0.92	0.99
Montana	0.19	0.18	0.21	0.32	0.78	0.89	1.38
Nebraska	0.56	0.49	0.54	0.54	0.77	0.82	0.83
Nevada	0.36	0.59	0.43	0.39	0.43	0.50	0.44
New Hampshire	0.45	0.85	0.53	0.53	0.95	0.64	0.65
New Jersey	1.19	1.08	1.19	1.16	1.51	1.58	1.55
New Mexico	0.38	0.34	0.33	0.38	0.46	0.41	0.46
New York	0.82	0.73	0.87	0.85	0.91	0.97	0.94
North Carolina	0.58	0.37	0.49	0.45	0.47	0.60	0.55
North Dakota	0.38	0.35	0.33	0.30	0.38	0.71	0.63
Ohio	0.90	0.67	0.71	0.69	0.79	0.82	0.80
Oklahoma	0.31	0.37	0.32	0.34	0.47	0.43	0.46
Oregon	0.50	0.49	0.44	0.47	0.55	0.50	0.54
Pennsylvania	0.85	0.82	0.82	0.79	1.06	1.23	1.18
Rhode Island	0.47	0.97	0.70	0.64	1.21	1.01	0.92
South Carolina	0.43	0.51	0.40	0.46	0.55	0.45	0.51
South Dakota	0.39	0.28	0.58	0.49	0.42	0.74	0.62
Tennessee	0.69	0.66	0.60	0.62	0.87	0.89	0.92
Texas	0.89	0.68	0.67	0.65	1.02	1.00	0.98
Utah	0.38	0.43	0.40	0.43	0.73	0.64	0.69
Vermont	0.57	0.49	0.53	0.54	0.60	0.65	0.67
Virginia	0.46	0.37	0.42	0.41	0.58	0.43	0.43
Washington	0.60	0.72	0.70	0.68	0.86	0.83	0.80
West Virginia	1.41	1.05	1.37	1.73	2.13	2.61	3.29
Wisconsin	0.60	0.68	0.45	0.53	2.31	0.48	0.57
Wyoming	0.33	0.29	0.51	0.46	0.38	0.63	0.56
United States	0.59	0.55	0.55	0.56	0.82	0.78	0.80

[a]The total number of ELL students was obtained from the EDEN database.

[b]The number of tested, not proficient students was computed from the Consolidated State Performance Reports by subtracting the number of all LEP (ELL) students proficient or above on the state annual ELP assessment from the number of all LEP (ELL) students tested on the state annual ELP assessment.

[c]Counts for California for 2008-2009 were unavailable; percentage is based on the 2007-2008 count.

the ratios of ACS estimates and state-provided counts of all ELL students. The right-hand side of the table shows the ratios of ACS estimates and state-provided counts of tested not proficient students. Shown are the ratios of the ACS 1-year estimate to each school-year data from the state, as well as the ratios of the ACS 3-year estimate to the most recent school-year data from the state. Values that are more than 1.00 indicate that the ACS rate was higher than the state-provided rate. Values that are less than 1.00 indicate that the ACS rate was lower than the state-provided rate. The bottom row shows the ratio for the entire country, which provides a basis for comparing the ratios for each state.

The overall ratio of the ACS 3-year estimate to the state-provided estimate of all ELL students for 2008-2009 is 0.56, with a range from 0.27 for the District of Columbia to 1.73 for West Virginia. The overall ratio for the ACS 3-year estimate to the state-provided counts of tested, not proficient students for 2008-2009 is 0.80, with a range from 0.33 for Alaska to 3.29 for West Virginia. The lower rates for the ACS estimates than the state-provided counts lend some validity to the use of the "less than very well" criterion for ACS-based estimates (see Chapter 2), since setting the cut point lower would limit the ACS estimates to a much smaller ELL population.

UNDERSTANDING THE DIFFERENCES

We conducted a series of regression analyses to further examine the correspondence between the ACS estimates and the state-provided counts and to attempt to account for the differences. We were particularly interested in the extent to which the two rates tended to be proportional (i.e., differ by a consistent factor) or to deviate from proportionality. If the two rates are proportional, then the difference between them would have no effect on shares (although allocations could still be influenced by state practices, such as cut points.)

We conducted regression analyses to separately predict each year's worth of state data (2006-2007, 2007-2008, and 2008-2009), using the most recent 3-year ACS estimate as the explanatory variable. Correlations between state-provided rates and ACS rates were high,[1] reflecting the strong but imperfect association of rates that are based on the two data sources.

Figures 5-1 and 5-2 show a plot of the 3-year ACS rates (x-axis) and the state-provided rates (y-axis) by state, for the state-provided count of all ELLs and the state-provided count of tested not proficient students, respectively. Several states are clearly outliers on these plots. For instance, the state-based rate for California of all ELL students is nearly 25 percent, compared with an ACS estimate of only about 11 percent (see Figure 5-1). Nevada and New Mexico also have considerably higher rates on the basis of the state-provided counts than those based on the ACS estimates (see

[1]Correlations between the ACS 3-year rate and rates based on state-provided counts of total ELL students were 0.84 for 2006-2007, 0.86 for 2007-2008, and 0.90 for 2008-2009. Correlations between the ACS 3-year rate and rates based on state-provided counts of tested, not proficient students were 0.81 for 2007-2008 and 0.78 for 2008-2009.

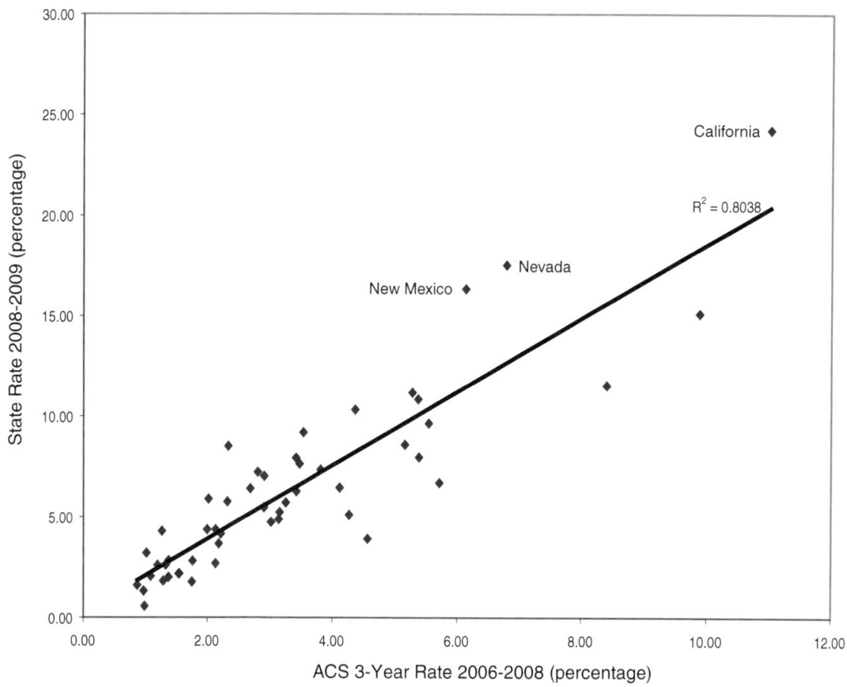

FIGURE 5-1 Comparison of ACS 3-year rate and state-provided rate of all ELL students for the 2008-2009 school year.

Figures 5-1 and 5-2). The rate based on the tested, not proficient count for Alaska is also considerably higher than the ACS estimate (see Figure 5-2).

We investigated several state-specific factors that might help to explain discrepancies between ACS estimates and state-provided counts, including demographic variables and testing practices.

Demographic Factors

We focused on demographic variables that represent characteristics of immigrant populations that might affect the differences between the state-provided counts and ACS estimates of ELL rates. We considered (1) the percentage of school-age immigrant children in the state, (2) the percentage of unauthorized immigrants in the state (Passel et al., 2006), (3) the percentage of unauthorized Mexicans in the state (Passel et al., 2006), (4) the percentage of foreign-born residents in the state with incomes below the poverty level, and (5) an indicator variable for states with large percentages of unauthorized immigrants (Arizona, California, New Mexico, Nevada, Texas).

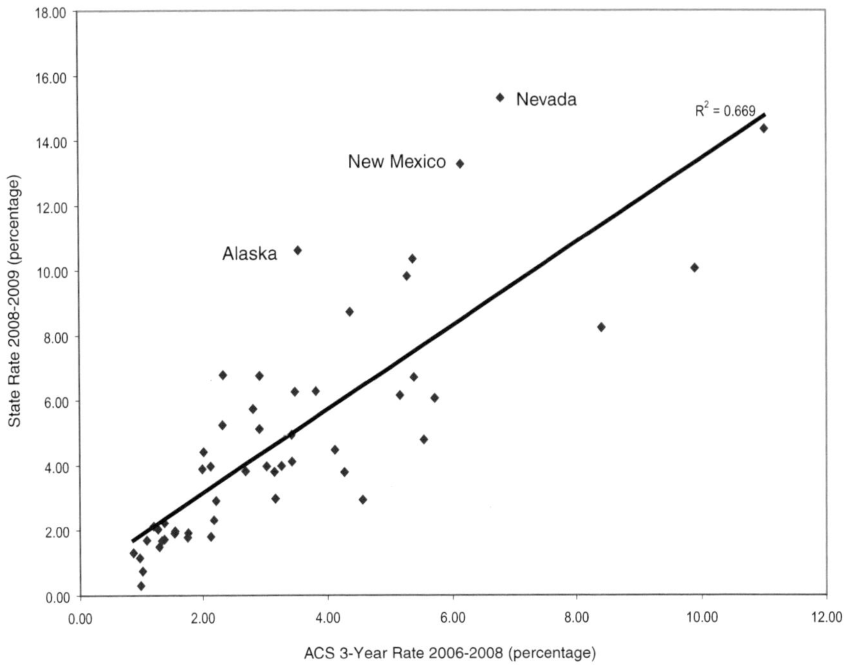

FIGURE 5-2 Comparison of ACS 3-year rate and state-provided rate of tested, not proficient students for the 2008-2009 school year.

These variables capture a number of hypothetical mechanisms. For example, unauthorized immigrants may be undercovered in the ACS although their school-age children would appear in the state counts. For another example, some groups of immigrant parents may overestimate the English speaking skills of their children relative to their own limited skills or by comparing them to requirements for community and family interactions rather than academic settings, particularly if their own formal education is limited.

State Practices

As discussed in Chapters 3 and 4, states use different tests, procedures, and criteria for classifying and reclassifying ELL students and for exiting them from programs. These differences might contribute to differential discrepancies between ACS estimates and state counts. It is difficult, however, to quantify the characteristics of state procedures, which are too complex and multidimensional to lend themselves to a simple rank ordering by stringency. Procedures also vary across districts in some states.

In the analyses shown in Tables 5-7 and 5-8, the dependent variable was the rate based on the state-provided count, and the ACS 3-year estimate was included as an explanatory variable, along with the other variables intended to test our hypotheses. Table 5-7 shows the results of the regressions for analyses in which the dependent variable was the rate based on the state-provided count of all ELL students. Table 5-8 shows the results when the dependent variable was based on the state-provided count of tested, not proficient students.

We ran nine different models for each year of state data (2007-2008 and 2008-2009). Model 1, the baseline model, included the 3-year ACS estimate as the only predictor of the state counts.[2] As shown in Table 5-7, the R^2 (the fraction of variance in state-provided rates explained by the predictor variable) for this model, when the rate of all ELL students was the dependent variable, was 0.73 for 2007-2008 and 0.80 for 2008-2009. Similarly, Table 5-8 shows that the R^2 for the basic model (Model 1), when the dependent variable was the rate of tested, not proficient students, was 0.66 for 2007-2008 and 0.66 for 2008-2009.[3]

Each of the successive models (2 through 9) added explanatory variables to examine the extent to which prediction of the state-provided counts might be improved. Models 2 through 7 included the variables intended to serve as proxies for the effects of the composition of the immigrant population in the state. Of these variables, only two were statistically significant predictors: the indicator variable for states (Arizona, California, New Mexico, Nevada, and Texas) with a high percentage of unauthorized immigrants (Table 5-7, Model 6, both years; Table 5-8, Model 6, 2007-2008) and the percentage of unauthorized Mexican immigrants (Table 5-8, Model 5, 2007-2008). The size and direction of the coefficients suggest that the ACS slightly underestimates the state-provided rate in states with high percentages of unauthorized immigrants (Table 5-7, Model 6, both years; Table 5-8, Model 6, 2007-2008). These findings provide some support for our hypothesis about the effects of the unauthorized immigrant population on the differences between the state counts and ACS estimates.

Models 8 and 9 both included indicators for states that used the ACCESS test (N, 18) or the ELDA test (N, 6). Neither of these indicators was a statistically significant predictor. We conducted one follow-up analysis, focused solely on the states that use the ACCESS test (the largest group in the country that uses the same test). As described in Chapter 3, states that use the ACCESS test are allowed to determine the specific proficiency level needed for ELL students to be exited from ELL classification. The ACCESS reports scores using six proficiency levels, and most states require students to score at least at the fourth level in order to be exited from the classification. However, some require students to score at the fifth level, some require students to score at the sixth level, and some specify additional criteria for the subtest scores. We attempted to classify states according to the stringency of their require-

---

[2]We have used a criterion of statistical significance of $p < .05$ (two sided).

[3]All R^2 values were adjusted for the sample size.

TABLE 5-7 Analysis of Using ACS 3-Year Estimate and Other Variables to Predict State-Provided Rate of All ELL Students

		2007-2008	2008-2009
Model 1	Adjusted R^2	.73**	.80**
	Regression Coefficients (Standard Error):		
	ACS 2006-2008	1.669** (.1449)	1.8305 (.1292)**
	Intercept	.0075 (.0054)	.0024 (.0052)
Model 2	Adjusted R^2	.73**	.79**
	Regression Coefficients (Standard Error):		
	ACS 2006-2008	1.6580 (.1949)**	1.9061 (.1754)**
	Percent immigrants	.0760 (.9245)	–.5845 (.9104)
	Intercept	.0073 (.0060)	.0042 (.0060)
Model 3	Adjusted R^2	.73**	.80**
	Regression Coefficients (Standard Error):		
	ACS 2006-2008	1.5330 (.4709)**	1.5345 (.4600)**
	Percent immigrants in poverty	.4413 (1.4583)	.9884 (1.4734)
	Intercept	.0077 (.0055)	.0028 (.0052)
Model 4	Adjusted R^2	.72**	.80**
	Regression Coefficients (Standard Error):		
	ACS 2006-2008	1.8110 (.3496)**	1.4574 (.2919)**
	Percent unauthorized immigrants	–.1358 (.5039)	.5911 (.4157)
	Intercept	.0060 (.0061)	.0040(.0053)
Model 5	Adjusted R^2	.74**	.82**
	Regression Coefficients (Standard Error):		
	ACS 2006-2008	1.4188 (.2432)**	1.4006 (.2118)**
	Percent unauthorized Mexicans	.7505 (.4617)	.9901 (.3973)*
	Intercept	.0096 (.0062)	.0075 (.0053)
Model 6	Adjusted R^2	.75**	.82**
	Regression Coefficients (Standard Error):		
	ACS 2006-2008	1.3943 (.2126)**	1.5047 (.1891)**
	States with high percent unauthorized	.0350 (.0161)*	.0323 (.0142)*
	Intercept	.0141 (.0068)*	.0102 (.0060)
Model 7	Adjusted R^2	.75**	.82**
	Regression Coefficients (Standard Error):		
	ACS 2006-2008	1.3241 (.2483)**	1.3282 (.2195)**
	Percent unauthorized Mexican	.3040 (.5449)	.7059 (.4622)
	States with high percent unauthorized	.0290 (.0194)	.0194 (.0163)
	Intercept	.0141 (.0068)*	.0107 (.0059)
Model 8	Adjusted R^2	.73**	.79**
	Regression Coefficients (Standard Error):		
	ACS 2006-2008	1.5814 (.1590)**	1.7979 (.1456)**
	ACCESS user	–.0078 (.0069)	–.0029 (.0069)
	ELDA user	–.0097 (.0090)	–.0042 (.0095)
	Intercept	.0140 (.0073)	.0052 (.0074)

TABLE 5-7 Continued

		2007-2008	2008-2009
Model 9	Adjusted R^2	.72**	.79**
	Regression Coefficients (Standard Error):		
	ACS 2006-2008	1.4786 (.4945)**	1.5912 (.4832)**
	Percent immigrants	–.1681 (.9603)	–.7492 (.9595)
	Percent immigrants in poverty	.4046 (1.4826)	.9851 (1.5130)
	ACCESS user	–.0801 (.0071)	–.0035 (.0070)
	ELDA user	–.0010 (.0095)	–.0056 (.0099)
	Intercept	.0148 (.0082)	.0084 (.0085)

NOTES: *$p < .05$; **$p < .01$. All R^2 values were adjusted for the sample size (n = 51).

ments on the ACCESS test. We then ran a regression on just these 18 states to see if the relationship between state counts and ACS estimates was improved when the stringency of ACCESS proficiency requirements was considered. The results did not support this hypothesis: no improvement in R^2 was evident when the stringency of proficiency requirements was considered.

We also ran a series of analyses to investigate the effects of changing the ACS cut score on the relationship between ACS estimates and state counts. For these analyses, we classified students as ELL students if their parents indicated their English speaking skills were "less than well" (rather than "less than very well"). We then examined the relationships between ACS estimates and state counts. The results showed no improvement in the relationships between ACS estimates and state counts: in fact, the percent of variance explained generally declined.[4] We also tried including ratios at all of the cut points in the same model: we found that those below the "less than very well" cut point were not significant predictors.

CONCLUSION 5-2 On the basis of the analysis of the effect of the American Community Survey (ACS) cut-point on the relationship between the ACS estimates and the state counts, the cut point of "less than very well"

[4]Two models were run. In the first, the dependent variable was the state provided count of **ELL students** divided by all public school enrollees for the 2007-2008 school year. The independent variable was the ACS 2006-2008 estimate of youths aged 5-18 years enrolled in public school enrolled who spoke English "less than well" divided by youths aged 5-18 years enrolled in public school. The adjusted R-square was 0.49 and the ACS variable was significant at 0.01 level. In the second model, the dependent variable was the state-provided count of **tested, not proficient ELL students** divided by all public school students enrolled for the 2007-2008 school year. The independent variable was the ACS 2006-2008 estimate of youths aged 5-18 years enrolled in public school who spoke English "less than well" divided by 5-18 years old and enrolled in public school. The adjusted R-square was 0.38, and the ACS variable was significant at the 1% level. Thus, even though the "less than well" variable was significant, its t-value was much lower than the "less than very well" variable in Model 1 in Tables 5-7 and 5-8. Also, both of these models had much lower explanatory power relative to the model with "less than very well" as the explanatory variable.

TABLE 5-8 Analysis of Using ACS 3-Year Estimate and Other Variables to Predict State-Provided Rate of Tested, Not Proficient ELL Students

		2007-2008	2008-2009
Model 1	Adjusted R^2	.66**	.66**
	Regression Coefficients (Standard Error):		
	ACS 2006-2008	1.300** (.1300)	1.2871 (.1293)**
	Intercept	.0034 (.0052)	.0057 (.0051)
Model 2	Adjusted R^2	.66**	.66**
	Regression Coefficients (Standard Error):		
	ACS 2006-2008	1.320 (.1773)**	1.2752 (.1763)**
	Percent immigrants	–0.1566 (.920)	.0918 (.9152)
	Intercept	.0039 (.0060)	.0052 (.0062)
Model 3	Adjusted R^2	.67**	.66*
	Regression Coefficients (Standard Error):		
	ACS 2006-2008	0.8173 (.4597)	1.0598 (.4614)*
	Percent immigrants in poverty	1.612 (1.4722)	.7589 (1.4778)
	Intercept	.0040 (.0052)	.0060 (.0053)
Model 4	Adjusted R^2	.67**	.66*
	Regression Coefficients (Standard Error):		
	ACS 2006-2008	0.9594 (.2950)**	0.9758(.2941)**
	Percent unauthorized immigrants	0.5396 (.4201)	.4930 (.4188)
	Intercept	.0049 (.0053)	.0071(.0053)
Model 5	Adjusted R^2	.71**	.68**
	Regression Coefficients (Standard Error):		
	ACS 2006-2008	0.8221 (.2102)**	0.9773 (.2185)**
	Percent unauthorized Mexicans	1.1009 (.3943)**	.7134 (.4099)
	Intercept	.0090 (.0053)*	.0094 (.0055)
Model 6	Adjusted R^2	.70**	.67**
	Regression Coefficients (Standard Error):		
	ACS 2006-2008	0.9105 (.1862)**	1.0807 (.1954)**
	States with high percent unauthorized	.0387 (.0140)**	.0205 (.0146)
	Intercept	.0127 (.0059)*	.0107 (.0062)*
Model 7	Adjusted R^2	.71**	.67**
	Regression Coefficients (Standard Error):		
	ACS 2006-2008	0.7279 (.2154)**	0.9395 (.2290)**
	Percent unauthorized Mexican	.7306 (.4537)	.5649 (.4822)
	States with high percent unauthorized	.0253 (.0160)	.0102 (.0170)
	Intercept	.0132 (.0058)*	.0110 (.0062)
Model 8	Adjusted R^2	.66**	.66**
	Regression Coefficients (Standard Error):		
	ACS 2006-2008	1.2327(.1449)**	1.2208 (.1444)**
	ACCESS user	–.0080(.0068)	–.0067 (.0067)
	ELDA user	–.0056 (.0095)	–.0074 (.0094)
	Intercept	.0093 (.0074)	.0114 (.0074)

TABLE 5-8 Continued

		2007-2008	2008-2009
Model 9	Adjusted R^2	.65**	.64**
	Regression Coefficients (Standard Error):		
	ACS 2006-2008	0.7988 (.4795)	1.0263 (.4833)*
	Percent immigrants	–.3598 (.9523)	–.1165 (.9598)
	Percent immigrants in poverty	1.5989 (1.5016)	.6992 (1.5134)
	ACCESS user	–.0083 (.0070)	–.0068 (.0070)
	ELDA user	–.0057 (.0098)	–.0074 (.0099)
	Intercept	.0111 (.0084)	.0120 (.0084)

NOTES: *$p < .05$; **$p < .01$. All R^2 values were adjusted for the sample size (n = 51).

on the ACS language item appears to best approximate school assessments of English language learner status.

It should be noted that the small number of units (50 states and the District of Columbia) limited the number of variables that could be considered. Furthermore, the concentration of ELL populations in relatively few states further limited our ability to infer systematic correlates of the discrepancies between the data sources.

Within-State Analyses

We also conducted a series of regression analyses to examine the relationships between ACS estimates and state-provided counts for school districts (local education agencies or LEAs) within each state. The purpose of this analysis was to assess how well the ACS and state-provided numbers tracked each other under a consistent set of procedures, criteria, and tests, that is, those of a single state. We obtained the 3-year (2006-2008) ACS estimates and the state-provided (2007-2008) counts of all ELL students for each unified school district for which they were available, which limited us to school districts with total populations of at least 20,000 (due to ACS release restrictions for small areas). Also excluded were several states for which LEA-level data were unavailable (California, New Jersey, and South Dakota). This analysis could be conducted only with rates based on state counts of all ELL students, since LEA counts of tested, not proficient students were not available. We formed the rate for each district (that is, we divided each of the counts by the number of K-12 students enrolled in public schools in the state).

For smaller units of analyses, such as most school districts, the sampling variability of ACS estimates of rates is generally greater than for states. Simple sample correlations would be attenuated by this error, underestimating the strength of the underlying relationship between the ACS and state-provided measures. We therefore used hierarchical models that adjust for the sampling variability of the ACS data to estimate this relationship. For these analyses, the dependent variable was the ACS estimate of the school district's ELL rate, and the explanatory variable was the state-provided estimate of the district rate. (Making the ACS rate the dependent variable facilitated specification of a hierarchical model in which the difference between the ACS estimate and its linear prediction from the state-reported rate is modeled as the sum of two random effects, one for ACS sampling error with known variance, and one for the discrepancy between the ACS and the state rate with variance to be estimated.)

Table 5-9 shows descriptive information for each of the states included in the analysis. The first two columns show the school enrollment and the number of unified districts in the state. The third column shows the overall rate of ELL students in the state based on state-provided information. The next three columns provide distributional information about the LEA rates within the state (based on the state-provided information): the average rate across the districts and the 20th and 80th percentiles of the LEA rates in the state. The seventh column shows the overall rate of ELL students in the state based on the ACS information. The eighth column presents the ratio of the ACS rate to the state-provided rate. The final column shows the sample correlation of the rates based on ACS estimates and state-provided counts for the unified school districts within a state. For instance, the correlation between the two sets of rates for the 58 unified school districts in Alabama was 0.697. The correlation is labeled "unadjusted" because it has not been corrected for sampling error associated with the ACS estimates.

Table 5-10 presents the results of the within-state regressions in states with at least 10 eligible LEA units, incorporating a correction for sampling error in the ACS estimates. The first four columns show the results from regressions that include the intercept in the model. The first two columns show, respectively, the regression coefficients for the intercept and for the rate based on the state-provided estimate. The third column shows the root mean square residual error (RMSE) of the model, which quantifies the amount by which the ACS estimates by LEA vary around the regression line. The fourth column shows the correlations after adjustment for sampling error. The median of these estimated correlation coefficients is 0.949, and the coefficient exceeds 0.90 in 30 of 41 states, although there are also a few states for which these LEA-level correlations are relatively low.

The fifth and sixth columns show parallel results (regression coefficients and RMSE) from the regressions that did not include the intercept in the model. The final column is the ratio of the errors from the two models (with and without intercepts). This ratio is usually not far from 1.0 (except in a few states where the denominator is very small due to an extremely good model fit), suggesting that the no-intercept (proportional) model fits the data almost as well as the unconstrained linear model. As noted previously, the proportional model implies that ACS-based and state-data-based allocations would be equivalent.

In general, the results suggest very good consistency between the ACS and state-provided numbers within states. This greater consistency, relative to similar models fitted at the state level, might be attributed to two features of the within-

TABLE 5-9 Descriptive Summaries of LEA-Level Data on Rate of ELL Students, by State

	State Counts		State ELL Rates						
State	School Enrollment	Number of LEAs	Overall Rate (%)	Mean of LEAs (%)	20%-tile (%)	80%-tile (%)	ACS Overall Rate (%)	Ratio of ACS/State	Unadjusted Correlation
Alabama	579,913	58	3.0	2.8	0.4	4.7	1.3	0.44	0.697
Alaska	93,838	5	7.4	5.8	2.4	10.6	2.6	0.34	0.207
Arizona	872,395	72	14.2	14.5	3.7	26.2	9.6	0.68	0.798
Arkansas	227,292	30	8.7	5.9	0.6	7.3	3.1	0.36	0.918
Colorado	684,657	35	11.3	11.9	2.5	21.4	5.8	0.51	0.920
Connecticut	378,744	56	7.3	5.3	1.4	10.1	3.9	0.54	0.877
Delaware	102,396	13	6.7	5.8	2.1	9.0	3.3	0.49	0.529
District of Columbia	57,877	1	7.1	7.1	7.1	7.1	2.3	0.33	NA
Florida	2,619,362	54	8.8	5.2	0.9	9.4	5.2	0.59	0.682
Georgia	1,487,247	97	5.2	3.5	0.7	5.4	3.3	0.64	0.885
Hawaii	179,897	1	10.4	10.4	10.4	10.4	4.4	0.42	NA
Idaho	180,200	20	5.2	5.5	0.3	11.3	2.4	0.46	0.904
Illinois	1,519,448	202	11.4	7.2	0.8	10.6	6.4	0.56	0.769
Indiana	705,862	87	5.5	4.9	1.0	7.5	2.4	0.44	0.776
Iowa	220,538	29	6.0	4.8	0.8	7.2	2.5	0.42	0.646
Kansas	272,573	28	9.6	9.2	1.6	12.9	4.1	0.42	0.907
Kentucky	463,556	54	2.4	1.4	0.2	1.8	1.6	0.67	0.647
Louisiana	606,547	49	1.8	1.2	0.1	1.6	1.4	0.76	0.418
Maine	45,917	12	5.8	4.3	0.4	6.3	3.7	0.63	0.797
Maryland	843,426	23	4.8	2.6	0.6	3.4	3.0	0.63	0.862
Massachusetts	639,309	110	8.1	4.8	0.6	9.3	5.3	0.65	0.819
Michigan	830,996	103	4.6	3.5	0.4	5.1	2.6	0.55	0.718
Minnesota	537,291	60	9.2	6.5	1.3	10.1	4.7	0.51	0.779
Mississippi	300,235	44	1.4	1.4	0.3	2.1	1.0	0.71	0.069

continued

TABLE 5-9 Continued

State	State Counts		State ELL Rates				ACS		
	School Enrollment	Number of LEAs	Overall Rate (%)	Mean of LEAs (%)	20%-tile (%)	80%-tile (%)	Overall Rate (%)	Ratio of ACS/State	Unadjusted Correlation
Missouri	551,434	65	2.7	2.2	0.5	2.9	2.1	0.79	0.671
Montana	58,413	15	2.3	1.9	0.2	3.5	1.3	0.54	0.290
Nebraska	169,074	15	8.6	7.2	1.4	11.8	4.0	0.47	0.781
Nevada	419,488	8	10.9	12.5	10.3	13.9	6.9	0.63	–0.550
New Hampshire	86,972	17	2.8	1.7	0.3	2.6	1.6	0.56	0.407
New Mexico	270,081	22	18.8	18.3	6.0	33.8	6.6	0.35	0.658
New York	1,063,809	175	4.9	4.2	0.8	7.1	3.1	0.64	0.746
North Carolina	1,373,592	89	9.1	8.0	2.9	12.4	3.5	0.39	0.826
North Dakota	44,279	6	4.2	3.6	0.3	7.0	1.9	0.45	0.782
Ohio	1,026,648	151	2.9	2.1	0.3	2.8	1.6	0.56	0.726
Oklahoma	325,060	30	8.4	5.6	2.3	8.1	2.7	0.32	0.717
Oregon	430,730	45	12.2	11.4	2.2	18.7	5.9	0.48	0.921
Pennsylvania	107,022	18	4.5	2.6	0.2	5.3	3.1	0.70	0.841
Rhode Island	112,975	19	5.3	3.0	0.6	3.7	4.9	0.92	0.801
South Carolina	649,424	53	4.2	3.5	1.2	5.1	2.1	0.50	0.685
Tennessee	848,210	70	2.9	1.8	0.3	3.0	1.9	0.65	0.733
Texas	3,770,908	207	16.4	12.3	4.4	19.0	11.0	0.67	0.901
Utah	513,430	19	8.7	9.2	3.1	16.1	3.4	0.39	0.950
Virginia	1,125,896	77	7.2	4.3	0.5	5.7	3.0	0.42	0.738
Washington	837,945	78	7.5	6.4	1.5	9.9	5.4	0.72	0.859
West Virginia	239,577	32	0.9	0.6	0.0	0.7	1.0	1.06	–0.163
Wisconsin	498,915	66	7.3	4.6	0.7	7.1	3.8	0.52	0.729
Wyoming	43,609	6	1.8	2.2	0.5	4.5	1.3	0.73	0.951

NOTE: Data include only eligible LEAs, as described in the text.

TABLE 5-10 Results of Within-State Regressions

	Model with Intercept, State Data Rate				No-Intercept Model			
	Intercept Coefficient	State Coefficient	RMSE	Adjusted Correlation	State Coefficient	RMSE	Number of LEAs	Ratio of RMSE Estimates
Alabama	0.0026	0.2508	0.0036	0.9040	0.3084	0.0033	58	0.93
Arizona	0.0216	0.3852	0.0316	0.8433	0.4765	0.0346	72	1.09
Arkansas	0.0049	0.2747	0.0003	0.9999	0.3099	0.0003	30	1.11
Colorado	0.0029	0.4477	0.0004	1.0000	0.4693	0.0013	35	3.20
Connecticut	0.0043	0.4038	0.0065	0.9596	0.4602	0.0067	56	1.03
Delaware	0.0171	0.1587	0.0027	0.8904	0.3816	0.0069	13	2.56
Florida	0.0073	0.4762	0.0127	0.8737	0.5540	0.0134	54	1.06
Georgia	0.0044	0.5006	0.0043	0.9814	0.5723	0.0048	97	1.13
Idaho	0.0053	0.3292	0.0005	0.9997	0.3958	0.0036	20	7.73
Illinois	0.0139	0.3808	0.0189	0.8834	0.4739	0.0220	202	1.17
Indiana	0.0034	0.2750	0.0035	0.9778	0.3224	0.0033	87	0.92
Iowa	0.0095	0.1688	0.0066	0.8514	0.2623	0.0096	29	1.46
Kansas	0.0017	0.3394	0.0102	0.9675	0.3508	0.0102	28	1.00
Kentucky	0.0046	0.4069	0.0033	0.9360	0.5251	0.0021	54	0.65
Louisiana	0.0061	0.2929	0.0020	0.9192	0.4801	0.0047	49	2.38
Maine	0.0037	0.3091	0.0054	0.9590	0.3769	0.0043	12	0.79
Maryland	0.0049	0.4777	0.0067	0.8872	0.5681	0.0074	23	1.11
Massachusetts	0.0064	0.4551	0.0042	0.9891	0.5173	0.0047	110	1.11
Michigan	0.0075	0.2692	0.0052	0.9318	0.3665	0.0068	103	1.31
Minnesota	0.0092	0.3170	0.0107	0.9038	0.3984	0.0123	60	1.16
Mississippi	0.0051	0.0874	0.0024	0.6004	0.2673	0.0001	44	0.04
Missouri	0.0068	0.3750	0.0019	0.9820	0.5429	0.0038	65	2.04
Montana	0.0045	0.1244	0.0013	0.8678	0.2576	0.0038	15	2.91
Nebraska	0.0081	0.3081	0.0006	0.9996	0.3758	0.0063	15	10.44

continued

TABLE 5-10 Continued

	Model with Intercept, State Data Rate				No-Intercept Model			
	Intercept Coefficient	State Coefficient	RMSE	Adjusted Correlation	State Coefficient	RMSE	Number of LEAs	Ratio of RMSE Estimates
New Hampshire	0.0060	0.3166	0.0004	0.9976	0.4780	0.0003	17	0.80
New Mexico	0.0118	0.1972	0.0271	0.7143	0.2418	0.0284	22	1.05
New York	0.0086	0.3293	0.0058	0.9533	0.4390	0.0078	175	1.33
North Carolina	0.0002	0.3369	0.0056	0.9625	0.3393	0.0056	89	1.00
Ohio	0.0057	0.2022	0.0028	0.9620	0.3289	0.0039	151	1.42
Oklahoma	–0.0014	0.2894	0.0045	0.9483	0.2735	0.0048	30	1.05
Oregon	0.0026	0.4207	0.0040	0.9963	0.4425	0.0021	45	0.52
Pennsylvania	0.0021	0.5318	0.0058	0.9485	0.5606	0.0058	18	1.00
Rhode Island	0.0019	0.7080	0.0022	0.9960	0.7493	0.0030	19	1.36
South Carolina	0.0047	0.2788	0.0035	0.9254	0.3800	0.0035	53	0.99
Tennessee	0.0030	0.3389	0.0037	0.8840	0.4274	0.0042	70	1.14
Texas	0.0018	0.6058	0.0197	0.9603	0.6152	0.0197	207	1.00
Utah	0.0020	0.3175	0.0077	0.9591	0.3352	0.0079	19	1.02
Virginia	0.0082	0.2299	0.0059	0.9420	0.2931	0.0084	77	1.42
Washington	0.0024	0.5805	0.0097	0.9715	0.6075	0.0095	78	0.98
West Virginia	0.0069	–0.0525	0.0020	0.2609	0.2488	0.0036	32	1.79
Wisconsin	0.0100	0.2911	0.0061	0.9139	0.4134	0.0070	66	1.15

NOTE: Data include only eligible LEAs, as described in the text.

state comparison: (1) the use of consistent procedures and criteria within most states but different ones in different states, and (2) the possibly greater similarity among immigrant populations within the same state than those in different states. The first of these reasons points to the difficulties in making present state-provided data comparable across states, while the second indicates possible difficulties in interstate comparability for ACS data. Nonetheless, the high degree of within-state consistency does give some reason for optimism that better consistency is achievable.

CONCLUSION 5-3 In the absence of other factors, such as the legislated minimum allocation, the American Community Survey and state-provided data would yield broadly similar allocations to most states. However, the differences in allocations to a few states are substantial and not readily explainable by such factors as region of the country, demographic characteristics of the English language learner population, or the proficiency test used by the state.

Temporal Variation

Another criterion for comparison of the ACS estimates and state counts is the degree of variation over time of the estimates for each state. There are conflicting values in consideration of such variation. *Responsiveness* refers to the tendency of a set of estimates to respond quickly to changes in conditions, such as rapid growth of the population of immigrant children in a state from one year to the next. This term suggests a positive value in that resources will be more rapidly directed to states with growing needs if a more responsive measure is used. *Volatility* refers to the tendency of estimates to vary or fluctuate from year to year. It suggests a negative value since such funding fluctuations make it more difficult to plan and maintain program continuity. Responsiveness contributes directly to volatility when populations are changing, but there are additional sources of volatility particular to each data source. Sampling variation contributes to purely random volatility in the ACS estimates. State data could become volatile when a state changes its tests, standards, or procedures from one year to the next or when there is an error or change in the mechanisms for reporting ELL counts from school districts to states to the DoEd.

Table 5-11 summarizes the volatility of ACS and state-provided estimates of ELL counts in two ways (parallel to those used in the sensitivity analyses in Chapter 2). The first is the sum of absolute changes in state shares, equivalent to twice the portion of the total allocation that would be moved from one state to another in consecutive years. The second is the mean absolute value of *relative* changes in shares, which summarizes the amount by which allocations in each state change relative to the size of its allocation. As expected, the single-year ACS changes are about equal in the two pairs of years (2006 to 2007 and 2007 to 2008). As explained in Chapter 2, the 3-year ACS estimates are much more stable, both because of the greater reliability of 3 years of data and because only one out of the years changes in overlapping

TABLE 5-11 Comparison of Volatility in ACS Estimates and State-Provided Counts (in percentage)

	Sum of Absolute Difference in State Shares	Mean Absolute Relative Difference in State Shares
ACS 2006 to ACS 2007		
All	**6.07**	**14.84**
Large	3.02	5.77
Medium	2.27	10.99
Small	0.71	21.16
Minimum	0.07	26.11
ACS 2007 to ACS 2008		
All	**6.47**	**12.84**
Large	3.38	5.64
Medium	2.46	11.02
Small	0.54	14.17
Minimum	0.09	26.72
ACS 2005-2007 to ACS 2006-2008		
All	**3.64**	**4.83**
Large	2.36	2.46
Medium	1.04	4.34
Small	0.21	5.99
Minimum	0.02	7.18
State 2006-2007 to State 2007-2008		
All	**10.55**	**11.90**
Large	6.81	7.79
Medium	3.21	11.80
Small	0.42	9.83
Minimum	0.11	23.57
State 2007-2008 to State 2008-2009		
All	**5.53**	**9.98**
Large	2.91	5.61
Medium	2.11	8.38
Small	0.42	13.01
Minimum	0.09	14.58

NOTES:
Large States: Arizona, California, Florida, Georgia, Illinois, New Jersey, New York, Texas, Washington.
Medium States: Colorado, Connecticut, Indiana, Kansas, Maryland, Massachusetts, Michigan, Minnesota, Missouri, North Carolina, New Mexico, Nevada, Pennsylvania, Oregon, Ohio, Oklahoma, South Carolina, Tennessee, Utah, Virginia, Wisconsin.
Small States: Alabama, Alaska, Arkansas, Delaware, Hawaii, Idaho, Iowa, Kentucky, Louisiana, Maine, Mississippi, Nebraska, New Hampshire, Rhode Island, West Virginia.
Minimum States: District of Columbia, Montana, North Dakota, South Dakota, Vermont, Wyoming.

3-year periods. (For the same reason, these estimates are also the least responsive.) Interestingly, the between-year changes in state-provided shares are much larger in 2006-2007 than in 2007-2008. We do not have enough detailed information about changes in state practices to identify specific reasons for the changes that might cause this variation and predict whether results would be similar in future years.

The more detailed information by state share grouping sheds more light on patterns of volatility. In absolute terms, the largest part of annual changes in share occurs in the states with relatively large shares; as noted above, these encompass about 74 percent of allocations. However in *relative* terms, these states show the least volatility by any measure. Since volatility in ACS estimates is largely driven by sampling variation, it is consistently larger in relative terms for each group of successively smaller states. The pattern is less consistent in the state-provided estimates, although generally the larger states tend to have more stable numbers. This stability may reflect the greater effects on smaller states of rapid changes in ELL population in a few local areas, or it may reflect changes in reporting. Overall, the 3-year ACS estimates appear to be the most stable, at the cost of some loss of responsiveness. And as discussed in Chapter 2, 1-year ACS estimates do not capture year-to-year changes with acceptable precision. The evidence is ambiguous on comparative stability of single-year ACS estimates and state-based estimates.

CONCLUSION 5-4 The superior precision and stability of the 3-year American Community Survey (ACS) estimates outweigh their slower responsiveness to changes and make them superior to the ACS 1-year estimates as a basis for allocations.

6

Comparability of Estimates of Immigrant School-Age Children

In this chapter we discuss the definitions of immigrant students from the two allowable data sources and compare those definitions. In discussing the American Community Survey (ACS), we apply several of the analytical techniques used in Chapter 2 when considering the English language learner (ELL) estimates. We also assess the strengths and weakness of estimates based on state administrative data.

As noted in Chapter 1, Title III of the Elementary and Secondary School Act requires the U.S. Department of Education (DoEd) to allocate funds to all 50 states, the District of Columbia, and Puerto Rico on the basis of a formula that incorporates the population of immigrant children and youth in each state. Specifically, the legislation states that 20 percent of the funds are to be based on the population of "recently immigrated children and youth (relative to national counts of these populations)." Section 3301(6) defines an "eligible immigrant student" as an individual who (A) is aged 3 through 21; (B) was not born in any state; and (C) has not been attending one or more schools in any one or more states for more than 3 full academic years. In this definition, language spoken by an individual is not a criterion for classification as an immigrant. This absence in the statutory definition may affect the allocation of the 20 percent of the funds to the extent that the English speaking ability of immigrant populations might vary by state because the source of the immigrant population varies by state.

As with the counts of eligible limited English proficiency (LEP) children and youth, the data on immigrant students can and have been derived from both the ACS and administratively reported state counts, and both data series have been used in the allocation formula. Prior to fiscal 2005, the DoEd allocated the immigration-related portion of Title III funds to the states on the basis of the state-reported counts of the number of immigrant children and youth; since then, the department has used ACS data for the number of immigrant children and youth to determine the state allocations.

ACS DATA AND ESTIMATES

Definition of Immigrant Children and Youth

Like the ACS data that the department uses to determine students with LEP, the data on immigrant status are based on self-reports. Three ACS questions are used to identify recent immigrants: (1) whether each household member was born in the United States, (2) whether he or she is a citizen, and (3) for those not born in the United States, when the person entered the country—see Box 6-1. Household members between the ages of 3 and 21 are classified as recent immigrants if they are not U.S. citizens at birth[1] and entered the country less than 3 years prior to the survey.

Evaluation of the Survey Questions

Like the questions on language spoken and English speaking ability, the ACS questions that define an immigrant child or youth were adopted from the long form of the decennial census at the time the ACS was developed. They have also been a part of the Current Population Survey for some time. They play a critical role in the Census Bureau's annual population estimates program as the basis for the net international migration estimate. Owing to their importance, the objectivity and collectability of these questions has been the subject of several analyses over the years, culminating in a major 2006 Census Bureau ACS test of the new and modified item content (Harris et al., 2007).

Although much of the research has focused on missing content, such as parental nativity and date of naturalization, the "year of arrival" question has been the subject of some evaluation because of the concern that the current question allows reporting of only one entry to the United States even when the respondents have entered multiple times, and the interpretation of "coming to live" in the United States may be too broad. Redstone and Massey (2003) identified problems with the year of entry question as a source of underestimation of the number of years that have elapsed since a person's arrival. Most likely, the inconsistencies were the result of multiple entries into the United States by persons who may have provided the year of a recent entry rather than their first entry (Schmidley and Robinson, 2003).

The content test report suggested that there was confusion among both respondents and ACS field representatives about the kind of information that the entry question was seeking and about how to report multiple arrivals. This confusion was evident in the content test itself, when a test group that was asked further probing questions about year of arrival was not able to provide accurate answers to the ques-

[1]The category includes respondents who indicate they are U.S. citizens by naturalization.

BOX 6-1
ACS Questions on Birth, Citizenship, and Year of Entry into the United States

1. Where was this person born?
□ In the United States
Print Name of State

□ Outside the United States
Print Name of Foreign Country, or Puerto Rico or Guam etc

2. Is this person a citizen of the United States?
□ Yes, born in the United States,
□ Yes, born in Puerto Rico, Guam, the U.S. Virgin Islands, or Northern Marianas
□ Yes, born abroad of U.S citizen parent or parents
□ Yes, U.S. citizen by naturalization
Print Year of Naturalization

□ No, not a U.S. citizen

3. When did this person come to live in the United States?
Print year of Arrival

tions. Based on the results of the content test, in 2008 the ACS added a question on year of naturalization.

Unfortunately, it is not known how these issues with the precise timing of the date of entry affect the precision of the estimate of immigrant children and youth for purposes of Title III allocations. If a significant number of children and youth who had originally arrived 3 years ago or earlier reported a subsequent arrival because of confusion over the meaning of the question, there would be tendency for the count of recent immigrant children and youth from the ACS to be an overestimate.

Effect of Nonresponse on Data Quality

We next considered the possible effect of item nonresponse on the ACS estimates. The allocation (imputation) rates (described in Chapter 2) for the "place of birth" item were 7.0 percent in 2008, which is considered moderate, and the allocation rates for the "year of entry" items were also moderate, at 10.4 percent. However, the amount of imputation required has trended upwards from year to year for each of the immigrant-identifying questions: see Table 6-1.

The implications of nonresponse for the accuracy of estimates are not evident. Its effects depend, first of all, on the accuracy of the assumptions underlying the

TABLE 6-1 Allocation Rates for Nonresponse on Immigrant Items in the ACS, 2005-2008 (in percentage*)

Item	2008	2007	2006	2005
Place of Birth	7.0	5.5	4.8	4.7
Citizenship	2.5	1.8	1.6	1.6
Year of Entry:	10.4	8.5	7.9	6.9
Total Population Not Born in U.S.				

*The item allocation rates for year 2005 are for housing units only. The item allocation rates for 2006 to 2008 include housing units and group quarters populations.
SOURCE: From the Census Bureau Quality Measures Page, http://www.census.gov/acs/www/UseData/sse/ita/ita_def.htm [May 2010].

procedures used to impute the missing values, and, consequently, on whether they impute recent immigration at approximately the correct rate. Furthermore, even if estimates of the total number of recent immigrants are inaccurate, Title III allocations to states would only be affected if the errors are disproportionate across states. For 2008, the range in the imputation rates for "place of birth" and "year of entry" across states are fairly narrow with interquartile ranges (the area between the 25th and 75th percentiles) of only 1.4 percent and 3.2 percent, respectively. Given the moderate overall item nonresponse rates and the fairly limited range of rates among the states, the effects of nonresponse on the allocations are not likely to be substantial.

The ACS Estimates

The 1-year and 3-year estimates of immigrant children and youth for 2005-2008 were derived from special tabulations provided by the Census Bureau: they are shown in Table 6-2. Along with the number of immigrant children and youth aged 3-21 years, the table shows the corresponding standard errors and coefficients of variation.

The panel's conclusions regarding the characteristics of the 1- and 3-year estimates for ELL estimates (see Chapter 2) also apply to the ACS estimates of immigrant children. Because each 3-year estimate is based on three times the sample size of the 1-year estimates, the standard errors of the former are substantially lower. The shares of the states that are based on 1-year estimates do not fluctuate a great deal.

The variation is further dampened when shares are based on 3-year estimates because consecutive 3-year estimates include 2 overlapping years (in this case, 2006 and 2007) and so are more stable than 1-year estimates: see Table 6-3. However, the 1-year estimates respond more quickly to changes in economic and social characteristics than the 3-year estimates.

The percentage share of each state's estimate of immigrant children and youth is shown in Table 6-4.

TABLE 6-2 Number of Immigrant Children and Youth Aged 3-21, by State

	ACS 2005			ACS 2006			ACS 2007			ACS 2008		
State	Estimate	SE	CV	Estimate	SE	CV	Estimate	SE	CV	Estimate	SE	CV
Alabama	7,710	862	0.11	7,335	829	0.11	9,815	1,404	0.14	5,405	920	0.17
Alaska	965	431	0.45	1,765	487	0.28	2,555	728	0.28	1,095	433	0.40
Arizona	35,660	2,855	0.08	32,565	2,852	0.09	37,565	3,158	0.08	22,250	2,429	0.11
Arkansas	4,680	935	0.20	5,330	838	0.16	5,785	1,044	0.18	4,510	914	0.20
California	251,275	9,185	0.04	214,095	6,483	0.03	208,295	6,405	0.03	179,500	6,834	0.04
Colorado	16,835	1,897	0.11	13,405	1,539	0.11	13,560	1,558	0.11	12,780	1,417	0.11
Connecticut	10,670	1,395	0.13	11,275	1,573	0.14	11,095	1,550	0.14	10,165	1,073	0.11
Delaware	2,495	491	0.20	1,980	706	0.36	1,515	512	0.34	1,100	451	0.41
District of Columbia	1,285	454	0.35	4,065	810	0.20	1,445	456	0.32	1,635	362	0.22
Florida	93,535	4,263	0.05	88,770	4,553	0.05	83,245	3,785	0.05	69,790	5,551	0.08
Georgia	36,945	3,031	0.08	31,160	2,172	0.07	26,295	2,172	0.08	28,460	2,670	0.09
Hawaii	6,645	1,512	0.23	8,495	1,273	0.15	5,110	858	0.17	7,810	2,288	0.29
Idaho	5,010	1,343	0.27	3,550	837	0.24	4,130	679	0.16	4,425	732	0.17
Illinois	35,965	2,710	0.08	35,225	2,771	0.08	44,240	2,900	0.07	32,535	2,505	0.08
Indiana	11,985	1,368	0.11	11,160	1,436	0.13	10,665	1,015	0.10	9,160	1,172	0.13
Iowa	4,150	733	0.18	4,685	704	0.15	4,580	812	0.18	5,540	818	0.15
Kansas	6,035	913	0.15	7,030	1,236	0.18	6,815	940	0.14	6,330	988	0.16
Kentucky	5,275	961	0.18	5,600	679	0.12	6,355	1,202	0.19	5,710	961	0.17
Louisiana	3,185	602	0.19	4,980	798	0.16	5,940	1,113	0.19	4,910	910	0.19
Maine	995	497	0.50	1,715	469	0.27	1,075	393	0.37	1,520	436	0.29
Maryland	26,765	2,621	0.10	23,940	1,858	0.08	22,260	1,974	0.09	19,500	1,392	0.07
Massachusetts	23,935	1,876	0.08	21,920	1,855	0.08	24,380	2,174	0.09	26,290	2,327	0.09
Michigan	20,640	2,328	0.11	17,700	2,054	0.12	21,810	2,030	0.09	18,945	1,904	0.10
Minnesota	14,420	1,562	0.11	17,285	2,044	0.12	15,100	1,536	0.10	10,980	1,165	0.11
Mississippi	2,695	703	0.26	2,925	663	0.23	3,300	534	0.16	3,420	896	0.26
Missouri	7,315	1,244	0.17	7,370	1,123	0.15	8,090	1,062	0.13	8,940	1,031	0.12

continued

TABLE 6-2 Continued

	ACS 2005			ACS 2006			ACS 2007			ACS 2008		
State	Estimate	SE	CV	Estimate	SE	CV	Estimate	SE	CV	Estimate	SE	CV
Montana	465	170	0.37	1,240	366	0.30	1,415	462	0.33	800	281	0.35
Nebraska	4,130	904	0.22	4,820	759	0.16	2,790	560	0.20	3,885	797	0.21
Nevada	9,445	1,158	0.12	11,900	1,337	0.11	11,425	1,517	0.13	9,620	1,477	0.15
New Hampshire	1,155	368	0.32	2,635	779	0.30	1,640	439	0.27	1,395	381	0.27
New Jersey	38,670	2,645	0.07	38,475	2,245	0.06	34,525	2,470	0.07	37,725	2,451	0.06
New Mexico	5,720	1,063	0.19	9,465	1,439	0.15	3,920	759	0.19	2,340	500	0.21
New York	83,310	3,942	0.05	95,185	4,360	0.05	79,390	3,826	0.05	84,055	3,810	0.05
North Carolina	27,890	3,614	0.13	27,175	2,541	0.09	24,495	2,165	0.09	21,660	1,887	0.09
North Dakota	415	212	0.51	1,805	501	0.28	1,265	343	0.27	1,105	426	0.39
Ohio	13,525	1,331	0.10	11,720	1,531	0.13	12,625	1,638	0.13	16,370	1,932	0.12
Oklahoma	5,935	1,021	0.17	6,515	1,025	0.16	7,950	1,123	0.14	5,305	902	0.17
Oregon	10,925	1,638	0.15	12,480	1,255	0.10	9,450	1,388	0.15	9,860	1,303	0.13
Pennsylvania	16,150	1,430	0.09	18,285	1,856	0.10	21,255	2,165	0.10	19,565	1,498	0.08
Rhode Island	4,610	1,169	0.25	3,465	712	0.21	3,410	798	0.23	3,205	636	0.20
South Carolina	11,865	1,452	0.12	7,005	903	0.13	9,715	1,200	0.12	5,695	845	0.15
South Dakota	1,835	876	0.48	715	241	0.34	665	185	0.28	180	176	0.98
Tennessee	9,800	1,260	0.13	10,845	1,152	0.11	11,885	1,333	0.11	10,150	1,670	0.16
Texas	130,990	5,851	0.04	122,375	5,277	0.04	110,375	5,088	0.05	95,575	4,515	0.05
Utah	7,410	1,179	0.16	7,950	1,048	0.13	9,420	1,115	0.12	8,630	1,569	0.18
Vermont	645	178	0.28	880	394	0.45	805	204	0.25	970	283	0.29
Virginia	25,835	2,306	0.09	26,545	2,072	0.08	23,800	1,673	0.07	22,240	1,691	0.08
Washington	24,375	2,018	0.08	28,775	3,064	0.11	31,535	2,641	0.08	24,160	2,083	0.09
West Virginia	200	108	0.54	945	252	0.27	1,335	318	0.24	2,095	573	0.27
Wisconsin	8,805	1,057	0.12	9,110	1,029	0.11	7,525	890	0.12	6,300	1,010	0.16
Wyoming	1,085	422	0.39	1,290	362	0.28	680	281	0.41	165	92	0.56
United States	1,082,255	17,490	0.02	1,046,930	14,440	0.01	1,008,330	13777	0.01	895,760	15,175	0.02

NOTES: SE = standard error; CV = coefficients of variation.
SOURCE: U.S. Census Bureau Special Tabulations.

TABLE 6-3 Average Number of Immigrant Children and Youth Aged 3-21, by State

	ACS 2005-2007			ACS 2006-2008		
State	Estimate	SE	CV	Estimate	SE	CV
Alabama	8,680	575	0.07	7,295	607	0.08
Alaska	2,080	483	0.23	1,800	303	0.17
Arizona	35,815	1,995	0.06	30,470	1,584	0.05
Arkansas	5,355	486	0.09	5,300	557	0.11
California	225,860	4,538	0.02	198,565	3,896	0.02
Colorado	14,790	942	0.06	13,305	787	0.06
Connecticut	11,480	938	0.08	11,040	794	0.07
Delaware	1,950	367	0.19	1,455	353	0.24
District of Columbia	2,360	312	0.13	2,165	260	0.12
Florida	89,035	2,463	0.03	80,605	2,367	0.03
Georgia	31,355	1,458	0.05	27,390	1,319	0.05
Hawaii	6,540	713	0.11	7,065	913	0.13
Idaho	4,715	793	0.17	4,195	527	0.13
Illinois	39,335	1,603	0.04	36,725	1,497	0.04
Indiana	11,545	733	0.06	9,980	605	0.06
Iowa	4,635	404	0.09	4,935	459	0.09
Kansas	6,740	547	0.08	6,340	623	0.10
Kentucky	6,370	584	0.09	5,785	505	0.09
Louisiana	4,820	674	0.14	5,230	578	0.11
Maine	1,345	293	0.22	1,650	353	0.21
Maryland	24,730	1,313	0.05	22,530	1,268	0.06
Massachusetts	24,605	1,118	0.05	24,085	1,312	0.05
Michigan	21,035	1,208	0.06	19,370	907	0.05
Minnesota	16,315	956	0.06	14,895	1,005	0.07
Mississippi	3,425	472	0.14	3,315	426	0.13
Missouri	8,280	696	0.08	8,230	678	0.08
Montana	1,060	190	0.18	1,305	228	0.17
Nebraska	4,075	477	0.12	3,785	422	0.11
Nevada	10,925	789	0.07	11,055	931	0.08
New Hampshire	1,995	281	0.14	1,860	319	0.17
New Jersey	36,740	1,357	0.04	36,920	1,386	0.04
New Mexico	6,105	588	0.10	5,265	557	0.11
New York	88,090	2,421	0.03	8,5340	2,635	0.03
North Carolina	25,720	1,382	0.05	23,915	1,160	0.05
North Dakota	1,455	270	0.19	1,300	200	0.15
Ohio	13,595	1,006	0.07	13,660	1,013	0.07
Oklahoma	7,345	622	0.08	6,920	687	0.10
Oregon	11,435	921	0.08	10,615	945	0.09
Pennsylvania	19,500	1,315	0.07	19,725	1,095	0.06
Rhode Island	3,920	564	0.14	3,350	483	0.14
South Carolina	9,950	810	0.08	7,910	713	0.09
South Dakota	1,155	300	0.26	540	135	0.25

TABLE 6-3 Continued

State	ACS 2005-2007 Estimate	SE	CV	ACS 2006-2008 Estimate	SE	CV
Tennessee	11,460	905	0.08	10,660	759	0.07
Texas	122,765	3,237	0.03	109,105	2,996	0.03
Utah	8,535	828	0.10	8,685	811	0.09
Vermont	745	138	0.19	800	160	0.20
Virginia	25,690	1,291	0.05	23,380	1,129	0.05
Washington	27,935	1,260	0.05	27,800	1,528	0.05
West Virginia	1,045	217	0.21	1,520	263	0.17
Wisconsin	8,615	673	0.08	7,340	518	0.07
Wyoming	1,005	223	0.22	750	164	0.22
United States	1,064,075	8853	0.01	977,220	9,638	0.01

NOTES: SE = standard error; CV = coefficients of variation.
SOURCE: U.S. Census Bureau Special Tabulations.

The corresponding standard errors are shown in Table 6-5. The percentage shares are the basis for the portion of the Title III allocations based on immigrants. The nine states with the largest shares—Arizona, California, Florida, Georgia, Illinois, New Jersey, New York, Texas, and Washington—together account for more than 70 percent of immigrant youth.

We also calculated the ratio of immigrant children and youth aged 5-18 to all children and youth within each state of that age who are enrolled in public school. This ratio is computed by simply dividing the ACS estimate of immigrant children and youth enrolled in public school by the ACS estimate of all children and youth in public school: see Table 6-6.

Tables 6-3 through 6-6 are based on special tabulations of ACS data provided by the U.S. Census Bureau.

Sensitivity of the Estimates to Variation in Subpopulations

We analyzed how the percentage share of states in the 3-year ACS estimates for 2006-2008 were affected by modifying the statutory criteria (3-21 years old, foreign born, and entered the United States after a particular year) to assess sensitivity of allocations to these criteria. We first limited the count to those aged 5-18 years old, an age range more similar to those reported to the states by local education authorities. We also examined the effect of limiting the count to either only those enrolled in school or only those enrolled in public schools.

Table 6-7 compares allocations with various combinations of these modifications to those under the base (statutory) criteria. Variations in age criteria did not influence the allocation of states very much (mean absolute difference, MAD, of 0.09%). For those aged 3-21, restricting the estimates to enrolled children and youth

TABLE 6-4 Share of Immigrant Children and Youth Aged 3-21, by State (in percentage)

State	ACS 2005	ACS 2006	ACS 2007	ACS 2008	ACS 2005-2007	ACS 2006-2008
Alabama	0.71	0.70	0.97	0.60	0.82	0.75
Alaska	0.09	0.17	0.25	0.12	0.20	0.18
Arizona	3.29	3.11	3.73	2.48	3.37	3.12
Arkansas	0.43	0.51	0.57	0.50	0.50	0.54
California	23.22	20.45	20.66	20.04	21.23	20.32
Colorado	1.56	1.28	1.34	1.43	1.39	1.36
Connecticut	0.99	1.08	1.10	1.13	1.08	1.13
Delaware	0.23	0.19	0.15	0.12	0.18	0.15
District of Columbia	0.12	0.39	0.14	0.18	0.22	0.22
Florida	8.64	8.48	8.26	7.79	8.37	8.25
Georgia	3.41	2.98	2.61	3.18	2.95	2.80
Hawaii	0.61	0.81	0.51	0.87	0.61	0.72
Idaho	0.46	0.34	0.41	0.49	0.44	0.43
Illinois	3.32	3.36	4.39	3.63	3.70	3.76
Indiana	1.11	1.07	1.06	1.02	1.08	1.02
Iowa	0.38	0.45	0.45	0.62	0.44	0.51
Kansas	0.56	0.67	0.68	0.71	0.63	0.65
Kentucky	0.49	0.53	0.63	0.64	0.60	0.59
Louisiana	0.29	0.48	0.59	0.55	0.45	0.54
Maine	0.09	0.16	0.11	0.17	0.13	0.17
Maryland	2.47	2.29	2.21	2.18	2.32	2.31
Massachusetts	2.21	2.09	2.42	2.93	2.31	2.46
Michigan	1.91	1.69	2.16	2.11	1.98	1.98
Minnesota	1.33	1.65	1.50	1.23	1.53	1.52
Mississippi	0.25	0.28	0.33	0.38	0.32	0.34
Missouri	0.68	0.70	0.80	1.00	0.78	0.84
Montana	0.04	0.12	0.14	0.09	0.10	0.13
Nebraska	0.38	0.46	0.28	0.43	0.38	0.39
Nevada	0.87	1.14	1.13	1.07	1.03	1.13
New Hampshire	0.11	0.25	0.16	0.16	0.19	0.19
New Jersey	3.57	3.68	3.42	4.21	3.45	3.78
New Mexico	0.53	0.90	0.39	0.26	0.57	0.54
New York	7.70	9.09	7.87	9.38	8.28	8.73
North Carolina	2.58	2.60	2.43	2.42	2.42	2.45
North Dakota	0.04	0.17	0.13	0.12	0.14	0.13
Ohio	1.25	1.12	1.25	1.83	1.28	1.40
Oklahoma	0.55	0.62	0.79	0.59	0.69	0.71
Oregon	1.01	1.19	0.94	1.10	1.07	1.09
Pennsylvania	1.49	1.75	2.11	2.18	1.83	2.02
Rhode Island	0.43	0.33	0.34	0.36	0.37	0.34
South Carolina	1.10	0.67	0.96	0.64	0.94	0.81
South Dakota	0.17	0.07	0.07	0.02	0.11	0.06
Tennessee	0.91	1.04	1.18	1.13	1.08	1.09
Texas	12.10	11.69	10.95	10.67	11.54	11.16
Utah	0.68	0.76	0.93	0.96	0.80	0.89
Vermont	0.06	0.08	0.08	0.11	0.07	0.08
Virginia	2.39	2.54	2.36	2.48	2.41	2.39
Washington	2.25	2.75	3.13	2.70	2.63	2.84
West Virginia	0.02	0.09	0.13	0.23	0.10	0.16
Wisconsin	0.81	0.87	0.75	0.70	0.81	0.75
Wyoming	0.10	0.12	0.07	0.02	0.09	0.08

SOURCE: U.S. Census Bureau Special Tabulations.

TABLE 6-5 Standard Errors of Shares of Immigrant Children and Youth Aged 3-21, by State (in percentage)

State	ACS 2005	ACS 2006	ACS 2007	ACS 2008	ACS 2005-2007	ACS 2006-2008
Alabama	0.08	0.08	0.14	0.10	0.05	0.06
Alaska	0.04	0.05	0.07	0.05	0.05	0.03
Arizona	0.26	0.27	0.31	0.27	0.19	0.16
Arkansas	0.09	0.08	0.10	0.10	0.05	0.06
California	0.76	0.55	0.57	0.68	0.39	0.34
Colorado	0.17	0.15	0.15	0.16	0.09	0.08
Connecticut	0.13	0.15	0.15	0.12	0.09	0.08
Delaware	0.05	0.07	0.05	0.05	0.03	0.04
District of Columbia	0.04	0.08	0.05	0.04	0.03	0.03
Florida	0.37	0.42	0.36	0.61	0.22	0.23
Georgia	0.27	0.20	0.21	0.29	0.13	0.13
Hawaii	0.14	0.12	0.08	0.25	0.07	0.09
Idaho	0.12	0.08	0.07	0.08	0.07	0.05
Illinois	0.24	0.26	0.28	0.27	0.15	0.15
Indiana	0.13	0.14	0.10	0.13	0.07	0.06
Iowa	0.07	0.07	0.08	0.09	0.04	0.05
Kansas	0.08	0.12	0.09	0.11	0.05	0.06
Kentucky	0.09	0.06	0.12	0.11	0.05	0.05
Louisiana	0.06	0.08	0.11	0.10	0.06	0.06
Maine	0.05	0.04	0.04	0.05	0.03	0.04
Maryland	0.24	0.17	0.19	0.15	0.12	0.13
Massachusetts	0.17	0.17	0.21	0.25	0.10	0.13
Michigan	0.21	0.19	0.20	0.21	0.11	0.09
Minnesota	0.14	0.19	0.15	0.13	0.09	0.10
Mississippi	0.06	0.06	0.05	0.10	0.04	0.04
Missouri	0.11	0.11	0.10	0.11	0.07	0.07
Montana	0.02	0.03	0.05	0.03	0.02	0.02
Nebraska	0.08	0.07	0.06	0.09	0.04	0.04
Nevada	0.11	0.13	0.15	0.16	0.07	0.09
New Hampshire	0.03	0.07	0.04	0.04	0.03	0.03
New Jersey	0.24	0.21	0.24	0.26	0.12	0.14
New Mexico	0.10	0.14	0.08	0.06	0.06	0.06
New York	0.34	0.40	0.36	0.39	0.22	0.26
North Carolina	0.33	0.24	0.21	0.21	0.13	0.12
North Dakota	0.02	0.05	0.03	0.05	0.03	0.02
Ohio	0.12	0.15	0.16	0.21	0.09	0.10
Oklahoma	0.09	0.10	0.11	0.10	0.06	0.07
Oregon	0.15	0.12	0.14	0.14	0.09	0.10
Pennsylvania	0.13	0.18	0.21	0.16	0.12	0.11
Rhode Island	0.11	0.07	0.08	0.07	0.05	0.05
South Carolina	0.13	0.09	0.12	0.09	0.08	0.07
South Dakota	0.08	0.02	0.02	0.02	0.03	0.01
Tennessee	0.12	0.11	0.13	0.19	0.08	0.08
Texas	0.50	0.48	0.48	0.47	0.29	0.29
Utah	0.11	0.10	0.11	0.17	0.08	0.08
Vermont	0.02	0.04	0.02	0.03	0.01	0.02
Virginia	0.21	0.19	0.16	0.18	0.12	0.11
Washington	0.18	0.29	0.26	0.23	0.12	0.15
West Virginia	0.01	0.02	0.03	0.06	0.02	0.03
Wisconsin	0.10	0.10	0.09	0.11	0.06	0.05
Wyoming	0.04	0.03	0.03	0.01	0.02	0.02

SOURCE: U.S. Census Bureau Special Tabulations.

TABLE 6-6 Ratio of Immigrant Children Aged 5-18 Enrolled in Public School to All Children Aged 5-18 Enrolled in Public School (in percentage)

State	ACS 2005	ACS 2006	ACS 2007	ACS 2008	ACS 2005-2007	ACS 2006-2008
Alabama	0.32	0.29	0.38	0.28	0.36	0.35
Alaska	0.20	0.88	1.01	0.58	0.87	0.88
Arizona	1.56	1.40	1.46	0.81	1.51	1.22
Arkansas	0.43	0.49	0.59	0.09	0.52	0.42
California	1.81	1.45	1.45	1.27	1.57	1.39
Colorado	0.90	0.86	0.72	0.67	0.86	0.74
Connecticut	0.86	1.04	0.98	0.77	0.97	0.95
Delaware	0.85	1.08	0.63	0.50	0.82	0.71
District of Columbia	0.90	1.93	1.19	0.40	1.28	1.08
Florida	1.77	1.58	1.52	1.29	1.64	1.46
Georgia	0.91	0.68	0.59	0.76	0.74	0.68
Hawaii	2.68	3.11	1.53	2.31	2.37	2.29
Idaho	1.18	0.77	0.62	0.87	0.94	0.80
Illinois	0.78	0.69	0.90	0.68	0.80	0.75
Indiana	0.48	0.43	0.36	0.35	0.41	0.36
Iowa	0.43	0.39	0.37	0.49	0.38	0.44
Kansas	0.84	0.58	0.53	0.50	0.67	0.54
Kentucky	0.42	0.32	0.38	0.43	0.39	0.38
Louisiana	0.21	0.16	0.32	0.30	0.23	0.23
Maine	0.13	0.50	0.17	0.25	0.27	0.40
Maryland	1.76	1.40	1.36	1.18	1.50	1.36
Massachusetts	1.21	0.95	1.21	1.09	1.14	1.08
Michigan	0.62	0.57	0.67	0.62	0.62	0.61
Minnesota	0.98	1.15	0.93	0.65	1.04	0.96
Mississippi	0.09	0.18	0.09	0.35	0.16	0.26
Missouri	0.38	0.38	0.31	0.40	0.38	0.38
Montana	0.12	0.52	0.21	0.21	0.34	0.45
Nebraska	0.35	0.66	0.47	0.65	0.54	0.57
Nevada	0.87	0.89	0.96	1.20	0.92	1.03
New Hampshire	0.31	0.55	0.18	0.19	0.36	0.28
New Jersey	1.22	1.26	1.22	1.52	1.23	1.37
New Mexico	0.74	1.29	0.72	0.19	0.89	0.72
New York	1.42	1.45	1.26	1.35	1.37	1.34
North Carolina	0.99	0.65	0.74	0.57	0.78	0.68
North Dakota	0.12	0.56	0.36	0.63	0.40	0.45
Ohio	0.44	0.30	0.28	0.37	0.36	0.32
Oklahoma	0.28	0.38	0.60	0.28	0.43	0.45
Oregon	0.77	0.79	0.61	0.58	0.76	0.65
Pennsylvania	0.46	0.44	0.59	0.41	0.49	0.47
Rhode Island	1.00	1.31	1.41	0.79	1.29	1.20
South Carolina	0.87	0.31	0.52	0.29	0.61	0.43
South Dakota	0.99	0.31	0.15	0.00	0.52	0.15
Tennessee	0.36	0.36	0.41	0.41	0.40	0.42
Texas	1.54	1.26	1.20	1.01	1.35	1.17
Utah	0.56	0.54	0.55	0.65	0.59	0.59
Vermont	0.09	0.66	0.28	0.42	0.28	0.37
Virginia	0.99	0.86	0.99	0.80	0.94	0.84
Washington	1.12	1.48	1.65	1.10	1.41	1.41
West Virginia	0.04	0.14	0.09	0.47	0.08	0.23
Wisconsin	0.47	0.50	0.28	0.32	0.42	0.36
Wyoming	0.15	0.69	0.20	0.13	0.40	0.36

SOURCE: U.S. Census Bureau Special Tabulations.

TABLE 6-7 Difference in the Percentage Share of Immigrants Aged 3-21 of States by Age Group, Enrollment Status, and Type of School

Modification to Base Criteria	Mean Absolute Difference in Share	Mean Absolute Relative Difference
Age Group Limited to 5-18 Years Old		
All	**0.09**	**8.26**
Large	0.23	3.70
Medium	0.08	7.05
Small	0.03	9.18
Minimum	0.02	17.02
Enrollment in School		
All	**0.16**	**12.46**
Large	0.43	7.05
Medium	0.17	10.73
Small	0.05	13.74
Minimum	0.03	23.45
Enrolled in Public School		
All	**0.10**	**11.84**
Large	0.19	5.45
Medium	0.12	8.68
Small	0.04	14.13
Minimum	0.04	26.74
5-18 Years Old and Enrolled in Public School		
All	**0.13**	**12.87**
Large	0.29	5.88
Medium	0.14	10.86
Small	0.07	17.56
Minimum	0.02	18.64

NOTES:
Large States: Arizona, California, Florida, Georgia, Illinois, New Jersey, New York, Texas, Washington.
Medium States: Colorado, Connecticut, Indiana, Kansas, Maryland, Massachusetts, Michigan, Minnesota, Missouri, North Carolina, New Mexico, Nevada, Pennsylvania, Oregon, Ohio, Oklahoma, South Carolina, Tennessee, Utah, Virginia, Wisconsin.
Small States: Alabama, Alaska, Arkansas, Delaware, Hawaii, Idaho, Iowa, Kentucky, Louisiana, Maine, Mississippi, Nebraska, New Hampshire, Rhode Island, West Virginia.
Minimum States: District of Columbia, Montana, North Dakota, South Dakota, Vermont, Wyoming.

causes a larger change in state allocations (MAD of 0.16 percent and MARD [mean absolute relative difference] of 12.46 percent). The estimates of immigrant children and youth are more sensitive to the restriction of enrolled status than the restriction of enrolled in public school. When the criterion is restricted to public school enrolled, the MAD is 0.10 percent and the MARD is 11.84 percent.

Reliability of Estimates of Change

We also assessed the reliability of estimates of change in the ACS ratio (as defined above), by assessing both interyear reliability of estimates of change and the significance of between-year changes in rates. An analysis of the interunit reliability of *changes* in ratios between consecutive years (following the procedures used to compute this measure in Chapter 2) yields results broadly similar to those in Chapter 2 for ELL estimates. Indeed, a global test of changes in ratio from 2006 to 2007 (using an ANOVA F-test) found that overall the changes were not significant ($F < 1$). Correspondingly, the estimates' variance components for interyear change in ratio was zero, indicating that the amount of change observed could be entirely explained by sampling variation. From 2007 to 2008, change was significant ($F = .72$, $p \approx 0.001$), but the reliability of the estimates of changes were generally small, at most 0.62 for any state and less than 0.50 for more than three-quarters of the states. Thus, across the range of state sizes, interyear changes were still largely confounded with sampling variation.

However, there were some outlying state changes in ratio that appeared statistically significant, as assessed by t-tests[2] of interyear changes: see Table 6-8. A total of 12 states (Alaska, Arizona, Arkansas, Hawaii, Mississippi, Montana, New Mexico, Oklahoma, South Carolina, Washington, West Virginia, and Wisconsin) had large and significant changes in their immigrant shares (based on 1-year estimates) relative to the previous year. Of these states, Montana is a "minimum" state whose allocation is unaffected by modest changes in immigrant population; Arizona and Washington are both "large" states and had large and significant changes in their immigrant share from 2007 to 2008; the other nine states that had big changes are "medium" or "small" states.

These large changes are most sensitively picked up by 1-year estimates. However, using those estimates gives great weight to interyear changes that in most states are mainly noise. This result supports use of the 3-year estimates on grounds of greater stability.

STATE PROCEDURES FOR DETERMINING IMMIGRANT STATUS

Although the state estimates of the immigrant population start with a standard definition that has been promulgated by the DoEd, the states take different pathways to identifying and reporting this group. The administrative record counts of immigrant students begin in the local school districts. The procedures for each state are established by the state education officials to ensure that reports received from the

[2]Generally, t-tests are the conventional tests for comparison of two independent sample means.

TABLE 6-8 Absolute Difference and Absolute Relative Difference in Ratio of Immigrant Children and Youth (in percentage)

State	ACS 2006 versus ACS 2005		ACS 2007 versus ACS 2006		ACS 2008 versus ACS 2007		ACS 2006-2008 versus ACS 2005-2007	
	Absolute Difference	Absolute Relative Difference	Absolute Difference	Absolute Relative Difference	Absolute Difference	Absolute Relative Difference	Absolute Difference	Absolute Relative Difference
Alabama	0.03	2.72	0.09	6.39	0.09	7.20	0.01	0.68
Alaska	**0.68**	**31.17**	0.13	3.48	0.44	13.78	0.01	0.24
Arizona	0.16	2.70	0.07	1.20	**0.66**	**14.51**	**0.29**	**5.23**
Arkansas	0.07	3.58	0.10	4.70	**0.50**	**36.38**	0.10	5.33
California	**0.35**	**5.40**	0.01	0.15	0.17	3.22	**0.19**	**3.13**
Colorado	0.04	1.08	0.14	4.51	0.04	1.56	0.12	3.67
Connecticut	0.18	4.75	0.06	1.53	0.21	5.90	0.02	0.54
Delaware	0.23	6.01	0.44	12.93	0.13	5.67	0.10	3.41
District of Columbia	1.03	18.17	0.74	11.92	0.79	24.91	0.20	4.18
Florida	0.19	2.89	0.06	1.01	0.23	4.05	**0.18**	**2.89**
Georgia	0.23	7.18	0.09	3.36	0.16	6.08	0.06	2.10
Hawaii	0.43	3.68	**1.58**	**17.05**	0.78	10.20	0.09	0.93
Idaho	0.41	10.50	0.15	5.37	0.25	8.27	0.14	4.17
Illinois	0.10	3.27	0.21	6.64	0.22	7.04	0.06	1.84
Indiana	0.05	2.55	0.07	4.36	0.01	0.53	0.04	2.70
Iowa	0.04	2.69	0.02	1.16	0.12	7.14	0.06	3.75
Kansas	0.26	9.03	0.05	2.40	0.02	1.17	0.12	5.15
Kentucky	0.10	6.85	0.06	4.48	0.05	3.04	0.01	0.51
Louisiana	0.05	6.96	0.17	17.52	0.03	2.11	0.01	0.84
Maine	0.37	28.88	0.34	25.10	0.08	10.02	0.13	9.37
Maryland	0.36	5.66	0.04	0.68	0.18	3.60	0.14	2.53
Massachusetts	0.27	6.14	0.26	6.03	0.11	2.43	0.06	1.36
Michigan	0.05	2.13	0.10	4.16	0.05	1.94	0.01	0.55

Minnesota	0.16	3.86	0.21	5.09	0.28	8.80	0.08	2.08
Mississippi	0.09	16.79	0.09	16.11	**0.26**	**29.27**	0.10	11.88
Missouri	0.00	0.16	0.07	4.89	0.09	6.16	0.00	0.09
Montana	**0.39**	**30.57**	0.31	21.02	0.00	0.06	0.11	6.72
Nebraska	0.31	15.14	0.19	8.25	0.17	7.62	0.04	1.60
Nevada	0.01	0.42	0.07	1.85	0.24	5.63	0.11	2.85
New Hampshire	0.24	13.82	0.37	24.97	0.01	0.89	0.08	6.03
New Jersey	0.05	0.92	0.05	0.95	0.31	5.58	0.13	2.58
New Mexico	0.56	13.75	0.57	14.22	**0.53**	**29.50**	0.16	5.11
New York	0.04	0.63	0.19	3.45	0.09	1.69	0.03	0.63
North Carolina	0.34	10.33	0.08	2.92	0.16	6.12	0.10	3.48
North Dakota	0.44	32.62	0.20	10.73	0.27	13.53	0.05	3.08
Ohio	0.14	9.54	0.02	1.49	0.08	6.55	0.04	2.93
Oklahoma	0.10	7.94	0.21	10.94	**0.31**	**17.92**	0.02	1.38
Oregon	0.02	0.58	0.18	6.58	0.03	1.12	0.11	3.79
Pennsylvania	0.02	1.32	0.16	7.59	0.19	9.26	0.02	0.87
Rhode Island	0.31	6.81	0.09	1.69	0.62	14.12	0.08	1.67
South Carolina	**0.56**	**23.78**	0.21	12.68	0.23	14.07	**0.18**	**8.54**
South Dakota	0.67	25.82	0.17	17.85	0.15	50.00	**0.38**	**28.03**
Tennessee	0.00	0.11	0.05	3.41	0.00	0.14	0.02	1.07
Texas	**0.28**	**4.98**	0.06	1.26	0.19	4.33	**0.18**	**3.58**
Utah	0.02	0.85	0.01	0.42	0.10	4.09	0.00	0.04
Vermont	0.58	38.31	0.39	20.49	0.14	10.23	0.09	6.84
Virginia	0.14	3.70	0.13	3.61	0.19	5.40	0.10	2.85
Washington	0.36	6.89	0.17	2.72	**0.55**	**10.05**	0.01	0.10
West Virginia	0.10	29.28	0.05	10.81	**0.38**	**34.06**	0.15	23.82
Wisconsin	0.03	1.38	**0.22**	**13.85**	0.04	3.16	0.06	4.08
Wyoming	0.54	32.24	0.49	27.46	0.07	10.25	0.04	2.93

NOTE: The percentages in boldface indicate large and significant changes relative to the previous year.

local school districts within the state are comparable.[3] In addition, as mentioned in Chapter 1, states are also required to use up to 15 percent of their Title III allotments for school districts with significant increases in school enrollment of immigrant children and youth.

Although all states use a common definition of immigrant students—students who were born outside the 50 states, the District of Columbia, Puerto Rico, and the Virgin Islands who are between 3 and 21 years of age and were enrolled for the first time in a U.S. school on or after a date 3 years ago—nuances in timing and reporting procedures may affect comparability from state to state. For example, in our review of state procedures we found that timing of the reference and submission periods varied between states. For example, the counts in California are the larger of counts that can be taken on October 7 or March 1, at the option of the local education authority (California Department of Education, 2010a). In Illinois, the counts are supposed to be taken in November and December and reported on January 15 (Illinois State Board of Education, 2009). And in New Jersey, the 2010 count was as of April, to be reported in May (New Jersey Department of Education, 2010).

The counts of immigrant children include both public and private school immigrant students. They are most often collected by local education agencies (LEAs), but, in at least one state, the data for private schools are to be forwarded directly from the private schools to the state education agency (SEA). Because of different reporting procedures, there is considerable uncertainty about the quality and coverage of both the public and private counts, although there is some evidence that state agencies do attempt to standardize the reporting. An example of the effort taken to assure the integrity of these estimates is the form with instructions provided to the local school districts by the Illinois State Board of Education: see Figure 6-1.[4] Other states have similar structured processes for determining the counts for immigrant education plans.

In sum, the panel agrees with the assessment of the 2006 U.S. Government Accountability Office (GAO) assessment that, despite documented attempts to standardize the reporting of immigrant student counts in the states, the GAO reported that, "with regard to data states collect on the number of children and youth who are recent immigrants, state officials expressed a lack of confidence in these data" (U.S. Government Accountability Office, 2006b, p. 23).

In part, the problem arose in the process of initially identifying immigrant students who met the definition. The GAO reported that its investigators were told by state officials in some of the 12 study states that "these data were not very reliable because school and school district officials did not ask about immigration

[3]Within-state comparability is important because there are monetary consequences associated with these counts under the Immigrant Education Program (IEP). The immigrant student counts determine the local education agency's eligibility for Title III IEP funds, and the amount of those funds that each local school district will receive.

[4]In 2009-2010, some $881,000 in Title III IEP funds were distributed in Illinois to local school districts based on the counts provided in the state report.

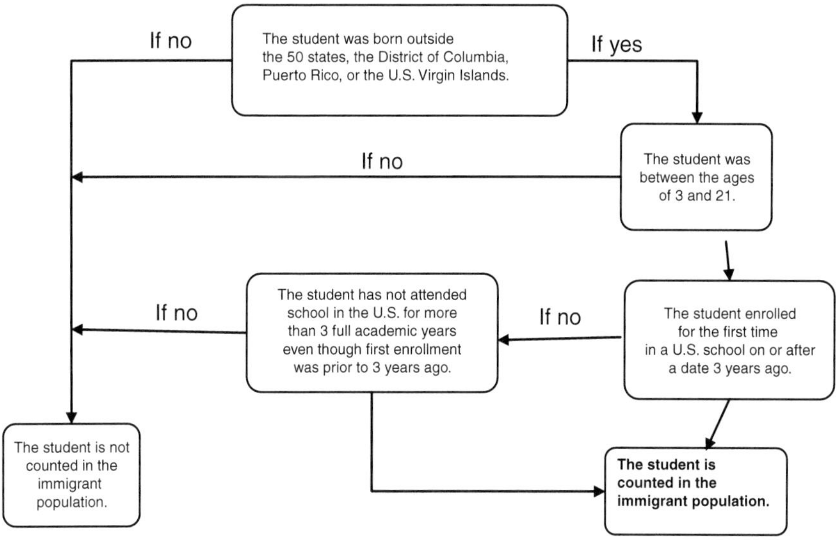

FIGURE 6-1 Rules for determining immigrant education program students in Illinois, 2009-2010 school year.
SOURCE: Illinois State Board of Education (2009, p. 3).

status directly" (U.S. Government Accountability Office, 2006b, p. 23). The GAO found that the school officials who were in charge of identifying recent immigrant students relied on such information as the student's place of birth and date of entry into the school system rather than taking steps to independently verify the student's status as a recent immigrant.

It is understandable that local school districts would be reluctant to pry too deeply into immigration status in order to verify eligibility as a recent immigrant student. The Supreme Court has ruled that public schools are prohibited from denying immigrant students access to public education on the basis of their immigrant status (*Plyler v. Doe*, 457 U.S. 202, 1982). Local school officials are encouraged, "when making inquiries for purposes of determining eligibility, to solicit voluntary information from parents and students or *finding alternative ways of identifying and documenting the eligibility of students*" (emphasis added) (New Jersey Department of Education, 2010, p. 3). The GAO report noted that officials in one state admitted that, in the absence of documentation, they assumed that students born outside the United States who entered the state's school system within the last 3 years were recent immigrants (U.S. Government Accountability Office, 2006b, p. 23).

COMPARISON OF ACS AND STATE ESTIMATES

There are important differences in the source, methodology, and results between the estimates of recent immigrant students that come from the ACS and the counts provided by the local school districts through the states. The differences in source and methodology were summarized in the GAO report: see Table 6-9. Not surprisingly, the very different sources and methods used in the two allowable sources result in very different estimates of the recent immigrant student population by state.

One way to depict the differences is to compute the ratio of the state student immigration counts with the ACS estimates of recent immigrant students. This comparison is shown in Table 6-10. In 17 states, the state count was higher than the ACS estimate, while it was lower in the remaining 34 states and the District of Columbia.

These relationships are illustrated graphically in Figure 6-2, which shows the ACS and state-reported counts for each state, together with the regression line through the origin (lower dashed line) and the line of equality (upper dashed line). Dotted lines around each state abbreviation represent 95 percent confidence intervals. Almost every state is below the line of equality, illustrating the generally lower

TABLE 6-9 Key Features of ACS and State-Collected Data on Immigrant Children and Youth

Feature	ACS Data[a]	State-Collected Data[b]
Measures Provided	Number of foreign-born persons aged 3 to 21 who arrived in the United States within the 3 years prior to the survey	Number of *(public and private school)* students in grades K-12 identified as recent immigrants
How It Is Measured	Self-report (sample of population)	States make determinations based on student records or other information. Some states told us that they are not able to directly ask students questions related to their immigration status.
Timing	Annual average of monthly sample	Varies
Purpose	To comply with Immigration Nationality Act and Public Health Service Act requirements; to provide data to set and evaluate immigration policies and laws.	To comply with the ESEA requirement to assess progress of all limited English proficient children, including immigrant children and youth, to attain English proficiency
(Department of) Education's Role in Data Collection	Work with Census to make sure appropriate questions are included. Can propose new questions, if necessary.	Education collects this number from the states in the Consolidated State Performance Reports.

[a]This column refers to data obtained by the U.S. Department of Education from ACS, but ACS collects additional data.

[b]Some states may have data available for children prior to kindergarten.

SOURCE: U.S. Government Accountability Office (2006b, p. 23).

TABLE 6-10 Comparison of State Student Immigrant Counts and American Community Survey Estimates of Recent Immigrant Students

State	ACS 3-Year Average, 2006-2008 Estimate	State 2007-2008 School Year Count	Ratio of ACS Estimate to State Count
Alabama	7,295	4,142	1.761
Alaska	1,800	880	2.045
Arizona	30,470	15,503	1.965
Arkansas	5,300	4,187	1.266
California	198,565	241,024	0.824
Colorado	13,305	12,940	1.028
Connecticut	11,040	13,571	0.813
Delaware	1,455	1,164	1.250
District of Columbia	2,165	993	2.180
Florida	80,605	142,333	0.566
Georgia	27,390	33,891	0.808
Hawaii	7,065	3,032	2.330
Idaho	4,195	3,188	1.316
Illinois	36,725	43,274	0.849
Indiana	9,980	11,763	0.848
Iowa	4,935	4,122	1.197
Kansas	6,340	11,206	0.566
Kentucky	5,785	7426	0.779
Louisiana	5,230	2,583	2.025
Maine	1,650	431	3.828
Maryland	22,530	16,617	1.356
Massachusetts	24,085	20,458	1.177
Michigan	19,370	11,052	1.753
Minnesota	14,895	15,985	0.932
Mississippi	3,315	6,007	0.552
Missouri	8,230	442	18.620
Montana	1,305	170	7.676
Nebraska	3,785	3,609	1.049
Nevada	11,055	14,694	0.752
New Hampshire	1,860	1,769	1.051
New Jersey	36,920	36,614	1.008
New Mexico	5,265	11,606	0.454
New York	85,340	98,797	0.864
North Carolina	23,915	23,365	1.024
North Dakota	1,300	497	2.616
Ohio	13,660	11,309	1.208
Oklahoma	6,920	4,954	1.397
Oregon	10,615	2,397	4.428
Pennsylvania	19,725	11,387	1.732
Rhode Island	3,350	2,903	1.154
South Carolina	7,910	6,415	1.233
South Dakota	540	197	2.741

continued

TABLE 6-10 Continued

State	ACS 3-Year Average, 2006-2008 Estimate	State 2007-2008 School Year Count	Ratio of ACS Estimate to State Count
Tennessee	10,660	1,5815	0.674
Texas	109,105	93,627	1.165
Utah	8,685	7,935	1.095
Vermont	800	556	1.439
Virginia	23,380	29,284	0.798
Washington	27,800	15,142	1.836
West Virginia	1,520	1,599	0.951
Wisconsin	7,340	5,882	1.248
Wyoming	750	391	1.918
United States	977,220	1,029,128	0.950

SOURCE: ACS estimates from U.S. Census Bureau Special Tabulation. State counts from U.S. Department of Education's EDEN Database.

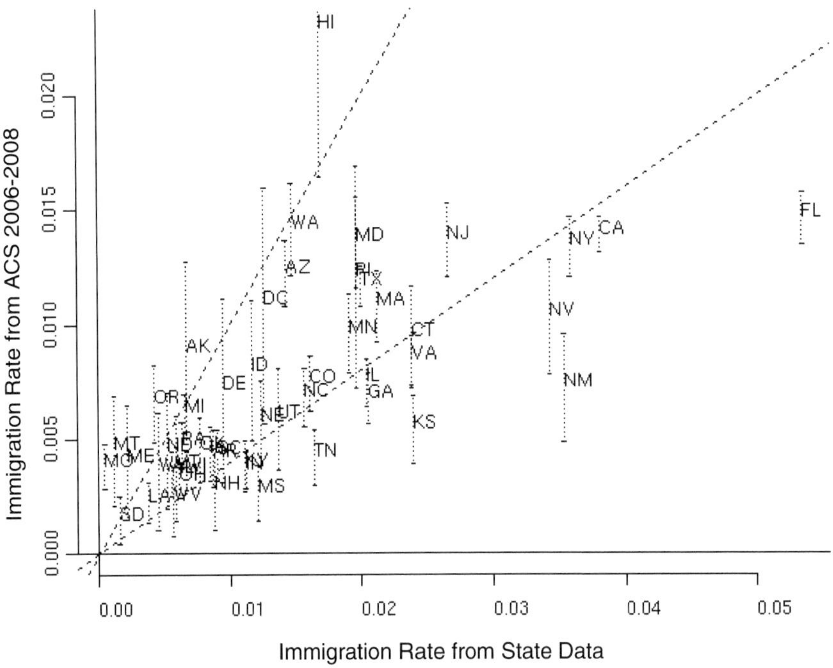

FIGURE 6-2 Immigrant ratio from state counts (2007-2008 academic year) and ACS 3-year estimates (2006-2008).

NOTES: The vertical dashed lines represent 95 percent confidence intervals for ACS estimates. The upper diagonal line is the line of equality; the lower diagonal line is proportional regression (regression through origin).

ACS estimates. States are scattered above and below the regression line, indicating the deviations from proportionality of ACS and state-reported estimates. However, only in some cases are these deviations from proportionality statistically significant (confidence interval does not cross regression line).

TABLE 6-11 Comparison of Volatility in ACS Estimates of Youth Aged 5-18 and Enrolled in Public School and State Counts of Recent Immigrants (in percentage)

	Sum of Absolute Difference in Share of States	Mean Absolute Relative Difference in Share of States
ACS 2006 to 2007		
All	**11.68**	**31.66**
Large	3.83	8.98
Medium	5.32	21.85
Small	2.05	42.43
Minimum	0.48	73.09
ACS 2007 to 2008		
All	**15.97**	**38.08**
Large	8.17	21.12
Medium	4.78	24.22
Small	2.75	53.58
Minimum	0.27	73.27
ACS 2005-2007 to 2006-2008		
All	**6.16**	**14.65**
Large	3.53	6.34
Medium	1.71	8.22
Small	0.73	20.90
Minimum	0.20	34.01
State 2006-2007 to 2007-2008		
All	**9.76**	**26.62**
Large	3.60	6.62
Medium	4.63	19.05
Small	0.93	20.11
Minimum	0.59	99.37
State 2007-2008 to 2008-2009		
All	**35.45**	**44.35**
Large	22.92	33.61
Medium	10.28	47.39
Small	2.02	41.39
Minimum	0.23	57.20

Volatility of State and ACS Estimates

As we did for the analysis of counts of ELL students (in Chapter 2), we report our assessment of changes in shares between consecutive years using several different measures of immigrant children, which are summarized in Table 6-11. Both the absolute and relative year-to-year changes in shares that are based on the 1-year ACS estimates are much greater than those that are based on the 3-year estimates. The 1-year estimates consequently are more volatile than the 3-year estimates, although they are also more responsive to year-to-year changes. In relative terms, the volatility of the ACS estimates increases monotonically from "large" to "small" states, as would be expected with diminished sample size, although in absolute terms the largest part of the share (and hence money moved) is in the "large" states.

The relative changes in share between consecutive 1-year ACS estimates are quite large, in most cases averaging more than 10 percent for all but the "large" states. This observation, with the earlier analysis of reliability of changes—which showed that interyear changes in these estimates are largely noise—together suggest that the volatility of the 1-year estimates outweighs the value of their greater responsiveness.

The interyear changes in shares based on state-provided data are surprisingly large, even though they are based on administrative data and therefore not subject to sampling error. This is especially notable from 2007-2008 to 2008-2009, when the shifts in share across states were even larger than those from single-year ACS estimates. (There seemed to be a substantial, but smaller, shift in single-year ACS estimates at about the same time.) This finding suggests that there might be substantial instability in the way child immigration counts are collected and reported by the states.

CONCLUSION 6-1 Due to greater stability and insensitivity to poorly estimated changes, the American Community Survey(ACS) 3-year estimates of immigrant children are statistically preferable to ACS 1-year estimates and more plausible at present than the state-provided counts.

ACS Data and LEA Reports

To better understand the relationship between ACS estimates and LEA-provided counts (through the states) of immigrant children, we performed an analysis of consistency between these measures across school districts within each state. The methods (correlations adjusted for sampling error and a hierarchical model), data (2006-2008 ACS), and inclusion criteria (districts with at least 20,000 population) are very similar to those presented in the parallel analysis of ELL rates in Chapter 5 and are not repeated here.

Table 6-12 presents summary statistics by state. Rates of immigrant status by school district vary substantially, often dramatically, within each state. ACS estimates are almost always lower than LEA-provided estimates, but the ratio varies greatly from state to state. This table shows that the state immigration rates were

TABLE 6-12 Rates of Immigrant Children by Eligible School District

	State Counts		State Immigrant Student Rates						
State	School Enrollment	Number of LEAs	Overall Rate	Mean of LEAs	20th Percentile	80th Percentile	ACS Overall Rate	Ratio of ACS/State	Unadjusted Correlation
Alaska	77,679	4	1.0	0.5	0.0	1.3	1.0	0.97	0.16
Arizona	834,896	64	1.7	1.5	0.1	3.0	1.4	0.82	0.50
Arkansas	217,450	28	1.6	1.1	0.1	1.7	0.6	0.38	0.60
California	5,491,668	411	4.1	3.9	2.0	5.6	1.4	0.33	0.51
Colorado	640,769	33	1.9	2.1	0.3	3.0	0.8	0.42	0.42
Connecticut	378,744	56	3.2	2.5	1.1	3.7	1.2	0.38	0.75
Delaware	102,396	13	1.1	0.9	0.5	1.1	0.7	0.62	0.09
District of Columbia	57,877	1	1.3	1.3	1.3	1.3	1.1	0.82	NA
Florida	2,619,362	54	5.4	3.6	1.2	5.8	1.5	0.27	0.51
Georgia	1,412,950	91	2.3	1.2	0.3	1.8	0.7	0.32	0.66
Hawaii	179,897	1	1.7	1.7	1.7	1.7	2.3	1.36	NA
Indiana	701,769	86	1.4	1.3	0.3	2.2	0.5	0.33	0.59
Iowa	210,815	27	1.3	1.0	0.1	1.9	0.6	0.46	0.39
Kansas	250,297	23	4.3	2.7	0.4	2.5	0.8	0.19	–0.05
Kentucky	371,406	38	1.9	0.8	0.1	0.7	0.6	0.30	0.45
Louisiana	566,824	43	0.4	0.3	0.1	0.4	0.3	0.58	0.40
Maine	26,010	6	1.5	1.4	0.0	3.3	1.8	1.18	0.92
Maryland	843,426	23	2.0	1.0	0.2	1.3	1.4	0.69	0.68
Massachusetts	590,965	97	3.2	2.4	0.2	4.2	1.4	0.45	0.64
Michigan	445,557	46	1.9	1.9	0.3	3.0	1.0	0.53	0.60
Mississippi	288,725	42	1.5	1.4	0.4	2.4	0.3	0.22	0.26
Missouri	94,131	9	0.4	0.5	0.2	0.7	0.8	1.74	–0.12
Montana	34,686	8	0.4	0.6	0.1	0.8	0.9	2.16	–0.63
Nebraska	169,074	15	1.7	1.4	0.4	2.5	0.8	0.47	0.27

continued

TABLE 6-12 Continued

	State Counts		State Immigrant Student Rates						
State	School Enrollment	Number of LEAs	Overall Rate	Mean of LEAs	20th Percentile	80th Percentile	ACS Overall Rate	Ratio of ACS/State	Unadjusted Correlation
Nevada	419,488	8	3.5	1.8	0.8	2.8	1.1	0.30	0.59
North Carolina	1,373,592	89	1.7	1.2	0.5	1.8	0.7	0.42	0.57
North Dakota	41,118	5	1.2	1.2	0.1	2.6	0.8	0.69	0.27
Oklahoma	274,584	23	1.5	1.0	0.3	1.8	0.7	0.46	0.33
Oregon	421,039	43	0.5	0.5	0.1	0.8	0.7	1.45	0.48
Pennsylvania	794,662	101	1.3	0.9	0.2	1.4	0.8	0.63	0.34
Rhode Island	112,975	19	2.2	1.1	0.2	1.5	1.4	0.63	0.72
South Carolina	632,088	49	1.0	0.8	0.2	1.3	0.5	0.47	0.54
Tennessee	806,634	61	2.0	0.9	0.1	1.5	0.5	0.23	0.30
Texas	3,616,573	183	2.4	1.7	0.6	2.5	1.3	0.56	0.60
Virginia	1,075,882	66	2.7	1.6	0.1	1.7	0.9	0.34	0.76
Washington	639,676	60	2.1	1.7	0.5	3.3	1.4	0.70	0.69
West Virginia	212,029	26	0.7	0.5	0.1	0.7	0.3	0.40	0.13
Wisconsin	413,481	49	1.1	0.8	0.3	1.2	0.5	0.43	0.36

NOTE: See text for discussion.

TABLE 6-13 Relationship at the School District Level Between ACS Estimates and State-Provided Estimates of the Rate of Immigrant Children Among Public School Enrollees, in Eligible Districts as Described in Text, for States with at Least 10 Eligible Districts

	Model with Intercept, State Data Rate				No-Intercept Model			Ratio of Sigma Estimates
	Intercept Coefficient	State Coefficient	Sigma	Adjusted Correlation	State Coefficient	Sigma	Number of LEAs	
Arizona	0.0071	0.1461	0.0035	0.6680	0.3562	0.0065	64	1.86
Arkansas	0.0028	0.1081	0.0004	0.9802	0.2268	0.0001	28	0.22
California	0.0033	0.1507	0.0036	0.7168	0.2217	0.0038	411	1.06
Colorado	0.0031	0.1506	0.0030	0.7115	0.2573	0.0032	33	1.05
Connecticut	–0.0015	0.3513	0.0011	0.9862	0.3003	0.0010	56	0.91
Delaware	0.0034	0.0030	0.0000	0.3748	0.2115	0.0001	13	1.17
Florida	0.0023	0.1699	0.0039	0.8306	0.2076	0.0038	54	0.98
Georgia	0.0007	0.2040	0.0016	0.8647	0.2328	0.0016	91	0.97
Indiana	0.0019	0.1100	0.0008	0.8801	0.1969	0.0011	86	1.43
Iowa	0.0015	0.1848	0.0007	0.9543	0.2636	0.0006	27	0.75
Kansas	0.0058	–0.0015	0.0005	0.1722	0.0271	0.0047	23	8.73
Kentucky	0.0021	0.1749	0.0001	0.9993	0.2432	0.0001	38	1.08
Louisiana	0.0003	0.2898	0.0003	0.9591	0.3554	0.0001	43	0.52
Maryland	0.0015	0.4868	0.0031	0.8727	0.5511	0.0032	23	1.01
Massachusetts	0.0031	0.2607	0.0015	0.9825	0.3320	0.0001	97	0.09
Michigan	0.0028	0.2796	0.0025	0.9454	0.3704	0.0025	46	1.00
Mississippi	0.0023	0.0936	0.0001	0.9956	0.1955	0.0001	42	1.05
Nebraska	0.0000	0.3624	0.0003	0.9965	0.3657	0.0003	15	0.85
North Carolina	–0.0003	0.2857	0.0013	0.9087	0.2686	0.0013	89	1.03
Oklahoma	0.0045	0.0144	0.0016	0.0836	0.2255	0.0028	23	1.76
Oregon	0.0023	0.8148	0.0015	0.9321	1.1533	0.0001	43	0.06
Pennsylvania	0.0041	0.0407	0.0020	0.1765	0.2971	0.0029	101	1.48

continued

TABLE 6-13 Continued

	Model with Intercept, State Data Rate				No-Intercept Model			Ratio of Sigma Estimates
	Intercept Coefficient	State Coefficient	Sigma	Adjusted Correlation	State Coefficient	Sigma	Number of LEAs	
Rhode Island	0.0016	0.4022	0.0002	0.9997	0.4494	0.0002	19	1.01
South Carolina	0.0023	0.1685	0.0015	0.5744	0.3782	0.0018	49	1.19
Tennessee	0.0020	0.0785	0.0012	0.6575	0.1358	0.0004	61	0.36
Texas	0.0019	0.3480	0.0031	0.8401	0.4218	0.0031	183	0.98
Virginia	0.0006	0.2743	0.0015	0.9836	0.2933	0.0014	66	0.94
Washington	0.0009	0.3521	0.0023	0.9128	0.3911	0.0027	60	1.15
West Virginia	0.0025	0.0736	0.0001	0.9938	0.2416	0.0001	26	1.12
Wisconsin	0.0024	0.0619	0.0012	0.4047	0.1859	0.0013	49	1.09

substantially different from the ACS rates in many states. In the most populous state, California, the state reports yielded an immigrant student estimate of 4.1 percent of all students in school year 2007-2008 while the ACS estimate was 1.4 percent. In most, but not all states, the state-reported rates were higher than the ACS rates.

Table 6-12 also summarizes the strength of the association between ACS and state-provided rates within each state, corrected for overfitting due to sampling error. With some exceptions, these correlations tend to be generally quite high, with half of the 30 states included showing adjusted correlations of higher than 0.90. This finding suggests that the measures are usually fairly consistent within each state, holding constant the state procedures and to some extent the immigration patterns (to the extent that they are more consistent within than between states). However, the correlations are considerably weaker in some states, perhaps providing evidence of inconsistent collection of immigration data or of varying patterns of immigration that affect consistency of reporting.

We also tested the relationship at the school district level between ACS estimates and state-provided estimates of rate of immigrant children among public school enrollees. As shown in Table 6-13, the results are mixed, with some states showing a very good consistency between the ACS and state-provided numbers for immigrants, and other states showing a very weak relationship between the series. When compared with the results of this test for the ELL estimates and counts in Chapter 5, these findings suggest that there are perhaps systemic differences between the ACS and state-provided counts at the school district level. The results suggest the possibility of less consistent procedures and criteria within many states than was observed with the within-state counts of ELL students, an indication that caution should be exercised in using the state-provided counts of immigrant children.

7

Decision Criteria and Recommendations

In this chapter we offer our recommendations for a set of criteria that the U.S. Department of Education (DoEd) can use in supporting a decision on which of the allowable data sources to use for allocating Title III funds. We present the decision criteria in the form of a list of desirable characteristics for data for formulas that allocate federal funds. Based on the analysis in Chapters 2 through 6, we relate this list of characteristics to the American Community Survey (ACS) estimates and the state-provided counts of English language learner (ELL) children and youth. Taking into account the decision criteria and mindful of the weight of evidence described in the analysis provided in this report, we also offer several recommendations to the DoEd and the U.S. Census Bureau with regard to the use of the allowable data sources for allocating Title III funds.

DESIRED CHARACTERISTICS OF ALLOCATION FORMULAS

It is useful to have objective criteria to assist in determining the appropriateness of a data source that is to be used in developing a formula for the allocation of federal funds. Fortunately, the committee has a starting point for assessing the two data sources that are allowable under Title III. Several considerations for evaluating the adequacy and appropriateness of data sources and their data elements for service in determining the allocation of federal funds were outlined in a National Research Council (NRC) report on formula allocations (National Research Council, 2001, p. 6): (1) the *conceptual fit* between currently available data and the formula elements, as defined in enabling legislation or administrative regulations; (2) the *level of geographic* detail for which data are provided; (3) the *timeliness* of the data; (4) the *quality* of the data; and (5) the *cost* of collecting or compiling new data to provide inputs to the formula. To these five considerations, we add five others that have emerged during our discussions with data users and producers and through our own examination of the necessary characteristics of data for use in allocating federal funds: (1) fairness, (2) stability from year to year, (3) insensitivity to differences in state policies and methods, (4) transparency, and (5) comparability. The rest of this section describes each of these criteria.

Conceptual Fit A data element used in an allocation formula should meet the conceptual objectives of the program for which the allocation is aimed. In the case of allocating Title III education funds to states, a data element with a good conceptual fit is one that meets the definition provided in the legislation—the number of limited English proficient and immigrant children and youth in a state. In a larger sense, however, considering the overall objective of the allocation of federal funds, a conceptually fitting data element would provide state and local governments with federal funding that is proportional to their need and circumstances.

Level of Geographic Detail The Title III legislation stipulates that the federal funds for the ELL program should be allocated to the states. Thus, the state government is the key level of detail for which the data should be available.

Timeliness The elapsed time between the reference period for the estimates and the period for which the allocations are being made should be as short as possible so that the allocation would appropriately reflect the need at the time that the allocation is made.

Quality Data quality is broadly defined as "fitness for use" (Statistics Canada, 2009, p. 6; Organisation for Economic Co-operation and Development, 2003, p. 6). In turn, fitness for use is generally characterized in terms of six attributes that are expected of the information provided by the data products:

1. *utility:* the usefulness of the information to its intended users;
2. *objectivity:* whether information is accurate, reliable, and unbiased, and is presented in an accurate, clear, and unbiased manner;
3. *interpretability:* the availability of documentation that includes a presentation of the underlying concepts and their definitions; descriptions of the methods used to collect, process, and analyze the data; and a discussion of the limitations imposed by the methods used to aid customers in understanding and using the data;
4. *integrity:* the security or protection of information from unauthorized access or revision;
5. *accuracy:* the difference between an estimate and its true value, characterized in terms of systematic error or bias, and random error or variance; and
6. *comparability:* similarity across geographic and demographic dimensions.

Of these, integrity, interpretability, and comparability are largely covered by other characteristics on our longer list, so under the heading of "quality" we focus on utility and accuracy.

Cost The benefits from improvements in conceptual fit or other aspects of data quality have to be weighed against the costs. Even when existing data sources are used, it is desirable to avoid incurring significant costs of obtaining data in a format suitable for the allocation process.

Fairness The allocation formula should be perceived as being fair. By fair, it is generally meant that the data used in the formula should be free from perverse manipulations, open to review, and should distribute resources equitably across governmental units. The formula itself should be replicable (see "Transparency," below).

Stability The data should be relatively stable over time. They should not be subject to extreme volatility or to large, unexplainable variation. However, in the context of Title III allocations, an appropriate balance needs to be struck between the stability of the data series and responsiveness to real annual changes in the size and characteristics of the ELL or immigrant population.

Insensitivity to Policies and Methodological Differences The data series should be relatively insensitive to differences that arise from administrative practices and policy differences between agencies and jurisdictions that provide the data and that benefit from those differences. In the context of allocations for Title III funding, if states X and Y have the same distribution of English language proficiency, but state X sets standards for program entry and exit that result in a larger fraction of its students designated as eligible for Title III services, the data series should take those differences into account so that they do affect the allocations. (This is not to suggest that states are or could be "gaming" the system. In fact, there is little incentive to game the system, since the cost to states and local education authorities of administering and conducting the ELL program generally exceeds the funds received from the federal government. In each year of the reauthorized Title III Program, the amount of money involved was small enough that it didn't create an incentive to states to this kind of strategic behavior.)

Transparency Users should be able to have access to and be able to understand the assumptions, methods, and results so that a knowledgeable user could readily reproduce the information, within the constraints of protecting the confidentiality and privacy of the subjects (see U.S. Census Bureau, 2010, p. 167).

Comparability The methodology by which the estimate is derived should be similar across geographical units.

The 2001 NRC report on this topic concluded that "there are many trade-offs among these considerations, and it is likely that no one data source will be superior to the others on all counts" (National Research Council, 2001, p. 6). The ability of the two allowable data sources to fulfill these desired characteristics is discussed in the next section.

COMPARING THE ALLOWABLE DATA SOURCES

In our judgment, both the ACS estimates and the state-provided counts meet each of these criteria to some extent, although each has strengths and weaknesses that need to be taken into account when considering their use for specific applications and at specific times. Table 7-1 shows our analysis of each of the 10 desired characteristics for each data source, discussed above. We present our ratings in the form of a scorecard, with the assignment of plus (+) marks, the highest rating being "++."

The panel particularly notes that prior issues with the relative volatility of the ACS data for smaller states have diminished as the survey has matured and as 3-year data have become available. Similarly, as discussed in Chapter 4, prior concerns about the accuracy and transparency of the state-provided data and the effects of different state practices on those data have abated over time as system-wide submission standards for data have been implemented. There are also some signs that states, as they are working to implement the guidelines of NCLB, are migrating toward more commonality in their approaches to identifying, testing, and educating the ELL population (as discussed in Chapters 1, 3, and 4.)

On the basis of within-state regression analysis (which has the effect of eliminating between-state differences), the panel found that, under a set of uniform state procedures, the ACS and state counts tracked very well. This result shows that the ACS serves as a relatively good proxy of the constructs addressed by the state English language proficiency (ELP) tests, and it has the advantage of being more uniform between the states. However, we could not demonstrate a similar relationship of the ACS to a more comprehensive ELP assessment at the state level because there is no measure of the latter that is uniform across states for such comparisons.

We found significant differences between the ACS and state-provided counts at the state level, and we attempted to explain these differences as a function of state policies. However, because state policies, procedures, and criteria differ along numerous dimensions that cannot be quantified in any parsimonious way, and the variables that described the states procedures and tests are so numerous, we were unable to identify any predominant cause of the differences in the state-level regressions reported in Chapter 5.

We find the conceptual fit of the state-provided counts to be particularly compelling in contrast to the ACS definition which is only a rough proxy for the official ELL definition. At the same time, we are concerned about the lack of state-to-state comparability in the policies, practices, and criteria for classifying students as English

TABLE 7-1 Comparison of ACS and State-Provided Data on Desired Characteristics for an Allocation Formula

Desired Characteristic	Evaluation	ACS	State Provided
Conceptual Fit	The ACS estimates define need in terms of the numbers of children and youth who are eligible for being served by virtue of their skill in speaking the English language. The state-provided counts define need in terms of the number of those identified by schools as being eligible by virtue of surveys and assessments that are becoming increasingly standardized. The state-provided data are considered to be more accurate and relevant assessments of individual students as well as of the intensity of need as defined by the policies of the various states.	+	++
Geographical Detail	The ACS estimates and the state-provided counts are available for both states and local education agencies (LEAs).	++	++
Timeliness	The ACS, state-level estimates for use in the allocation formula are available approximately 9 months following the reference period. The state-provided counts are submitted by the states to the Department of Education about 6 months after the school year data are collected in the fall and publicly released in July, which is also about 9 months after collection.	+	+
Quality	The data from the ACS meet statistical reliability standards as described in this report and are of acceptable precision. State-provided counts are based on administrative data and are not subject to sampling error, although there may be some different interpretation of the instructions for data collection. State-provided counts on immigrant children and youth very much rely on LEA judgments, they and fall short of the quality of the ELL counts or the ACS estimates.	++	+
Cost	Both the ACS estimates and state-provided counts of the ELL population are available at minimal extra cost.	+	+
Fairness	The Census Bureau has an excellent reputation for assuring that the data in its charge are free from manipulation. State data systems and submission procedures have improved such that the data are similarly free from manipulation, but states still have discretion over the timing of submissions and other policies that may affect perceptions of fairness.	++	+
Stability	The state-provided counts are relatively stable from year to year. The annual ACS estimates for smaller states have been subject to greater variation due to small sample sizes, but they are comparable. The 3-year estimates are more stable than both the 1-year ACS estimates and the state counts.	++	++

continued

TABLE 7-1 Continued

Desired Characteristic	Evaluation	ACS	State Provided
Insensitivity to Policy and Methodological Differences	The ACS estimates are not sensitive to administrative practices or policy differences, although they may be sensitive to differences in demographic composition of the respondents. The state-provided counts are somewhat sensitive to state decisions regarding identification, testing, and program entry and exit policies. The panel has no evidence that these state decisions are made in any way to influence the federal government's allocation of Title III funds. Nonetheless, the decisions would tend to influence the allocation.	++	+
Transparency	ACS data are collected by professional staff using highly standardized, well-documented methods. State data are collected by methods that vary from state to state and rely on implementation by local authorities; consequently, documentation of the methods as they are implemented across the country is not readily available.	++	+
Comparability	The ACS is comparable across geographic and demographic dimensions. The state-based counts conform to definitions promulgated by the U.S. Department of Education but are not comparable in their constructs due to differing state tests and classification and reclassification criteria.	++	+

language learners and reclassifying them as former English language learners that we have documented in Chapter 4.

We therefore believe that the DoEd should consider a new approach for the 80 percent of the funding that is based on the number of limited English proficient children in the state under Title III—one that uses both data sources, building on the strengths of each of them and recognizing their unique contributions. Because the panel concluded that the allocation formula would gain strength by using data from both data sources in the allocation formula, we discussed various means of doing so. We concluded that both data sources should be used in the allocation formula, with the predominant view that eventually the data sources should be given equal weight—balancing an emphasis on the current need in the state and an opportunity to dampen some of the variability in the ACS measure, and with an equal emphasis on a standardized measure across the states that the ACS offers. For the foreseeable future, the desirable characteristics of the ACS insofar as the quality of the data, the perception of fairness of the ACS, the insensitivity of the estimates to policy changes, and the comparability between geographic areas commend its continued use in the allocation formula.

However, as discussed above, the conceptual fit of the state-provided counts strongly commends their use in the allocation formula. Unfortunately, states' tests, practices, and procedures are not standardized across the country; and there is still much work to be done to improve the quality of the counts (see Chapters 3 and 4). Hence, we recommend a compromise approach that gives some weight initially to the state counts that can increase when the state data are judged to be of sufficient quality (considering the criteria enumerated in Table 7-1) for taking on the burden of constituting an equal share of the allocation. In evaluating these criteria, the DoEd might consider, as a matter of policy, whether the variation across the states in methods and ELP assessments are acceptable for use in determining allocations with the justification that they represent the procedures actually used to determine eligibility of students for services.The department also may wish to consider whether basing a component of the allocation estimates on counts based only on ELL students in public and charter schools is acceptable, given our results indicating that there may be a modest but not insignificant impact on allocations by limiting the counts to this population.

RECOMMENDATION 7-1 As soon as technically possible, the U.S. Department of Education should begin to incorporate state-provided counts of English language learner (ELL) students into Title III formula allocation calculations. Initially, the state-provided data should be given a weight of 25 percent of the ELL allocation, with the remaining 75 percent weight given to the American Community Survey data.

As discussed in Chapters 3 and 4, states annually report four sets of counts to the DoEd: the number of students who (1) are English language learners, (2) received Title III services, (3) took an ELP test, and (4) scored at the "English proficient" level on that test. Of these four counts, we recommend using those based on the results of the ELP test: that is, the count of ELL students who took an English proficiency test (3, above) and scored below the level the state defines as "English proficient (4, above)." We conclude that this count provides a relatively objective criterion within each state. Furthermore, in our judgment, this count is potentially the least likely to be affected by differing state policies and practices because even if nontest criteria vary (in specification and implementation) across states, the tests share a number of common features that provide a foundation for establishing comparability across them.

RECOMMENDATION 7-2 In the portion of the allocation that is based on state-provided data, the U.S. Department of Education should use the state-provided count of the number of students who are determined not to be English proficient on the basis of the state's English language proficiency test.

For the 20 percent of the allocation formula that is determined by the count of immigrant children, as discussed in Chapter 6, the committee does not find that the state-provided estimates have any significant benefits over those from the ACS.

RECOMMENDATION 7-3 The U.S. Department of Education should continue to use the American Community Survey estimate as the basis for allocating the 20 percent of the Title III funds that are to be based on the population of recently immigrated children and youth (relative to national counts of these populations).

This report has documented a number of areas in which both the ACS and the state-provided data would benefit from additional improvements. With regard to the ACS, we have highlighted the need for more research on the responses of the English speaking ability questions and the need to bring the ACS source questions used in defining the ELL population (currently based only on speaking ability) into closer alignment with professional and legal standards for determining limited English proficiency. Accordingly, we offer two recommendations about the ACS.

RECOMMENDATION 7-4 The U.S. Census Bureau should conduct research on the accuracy of the American Community Survey language item for assessing population prevalence of English language learner children and youth, including the strength of its association with more comprehensive English language proficiency (ELP) measures. With the objective of evaluating and improving the item, researchers should examine the effects on responses of situational, cultural, demographic, and socioeconomic factors, placement of the item in the questionnaire, and the ability of adult responders to make ELP distinctions.

With regard to the state-provided estimates, a program of research, evaluation, and enhanced data collection is likely to facilitate transition to more extensive use of state-provided data in funding allocations. This work should focus both on studies designed to improve the cross-state comparability of the performance levels that define which students are considered to be English proficient on an ELP test and on strategies for improving the quality of state-provided data.

There are several strategies the DoEd could consider for improving the cross-state comparability of state-provided data. One possibility would be to undertake quantitative studies to statistically link the ELP tests, although such studies are likely be of little value because many of the basic assumptions for the strongest form of linking (equating) have not been met. Although weaker linking methods with less stringent assumptions might be possible, we are not optimistic about their utility. Moreover, such studies would be resource intensive.

In contrast, qualitative approaches, such as the "crosswalk" analyses we described in Chapter 3, may be useful in evaluating the comparability of the performance levels. These studies could focus on the performance levels set by states to define "English proficient," seeking to evaluate the extent to which the skills required by the different states are comparable and determining a strategy for setting comparable performance levels across the states.

We also note that the current policy environment may foster increased comparability among the ELP tests. The current efforts directed at developing and adopting common English language arts and mathematics content standards across the states and collaborating to develop common assessment systems to measure these standards may affect the comparability of ELP tests. Specifically, changes to states' English language arts content standards are likely to trigger changes in states' ELP standards. We anticipate that as English language arts standards, math standards, and the language and literacy aspects of other content standards become more similar across states, so will states' ELP standards. As a result, the ELP tests that are used by states to measure these standards will likely become more similar and will more easily lend themselves to defining comparable cross-state performance standards for "English proficient."

With regard to improving the quality of state-provided data, there are several steps that the DoEd might explore. First, the department might consider asking states to provide documentation of the technical quality of their assessments, particularly information to document the procedures used to set the performance levels and to determine the "English proficient" level, as well as information to document the accuracy and validity of decisions based on the assessment. This type of information has been required for the English language arts and mathematics achievement tests used by the states to meet the accountability provisions of Title I of the Elementary and Secondary Education Act. These requirements might also be extended to the tests used for Title III, and they would help to ensure and enhance the quality of the data that states provide on their ELL students, particularly the composite and domain-performance levels used to define the "English proficient" standard.

Second, the DoEd could continue its efforts to improve the quality, consistency, and completeness of data collected from the states on the Consolidated State Performance Reports and maintained in the Education Data Exchange Network system.

As a strategy for improving the comparability of state-provided estimates of students who are considered to be English proficient on the basis of an ELP test becomes available and is implemented—and as evidence of the quality, consistency, and completeness of state-provided data improve—the state-provided data can be accorded more weight in the allocation formula.

RECOMMENDATION 7-5 When the quality and cross-state comparability of state-provided data have reached an acceptable standard, the weight given to the state-provided counts should be adjusted upward to the point at which the American Community Survey estimates and the state-provided counts contribute equally to the 80 percent portion of the allocation formula. State-provided counts should continue to be based on the number of students who are determined not to be English proficient on the basis of the state's English language proficiency test, in a way that is comparable across states.

References and Bibliography

Abedi, J. (Ed.). 2007. *English Language Proficiency Assessment in the Nation: Current Status and Future Practice.* Davis: University of California.

Abedi, J. 2008. Classification system for English language learners: Issues and recommendations. *Educational Measurement: Issues and Practice, 27*(3), 17-31.

Abedi, J., and Dietel, R. 2004. *Challenges in the No Child Left Behind Act for English Language Learners.* CRESST Policy Brief. Los Angeles: National Center for Research on Evaluation, Standards, and Student Testing, University of California.

American Educational Research Association, the American Psychological Association, and the National Council on Measurement in Education. 1999. *Standards for Educational and Psychological Testing.* Washington, DC: American Educational Research Association.

American Institutes for Research. 2005. *ELDA English Language Proficiency Standards and Test and Item Specifications for Grades 3–12.* Available: http://arkansased.org/educators/pdf/lep/elda_standards_specs_103105.pdf [January 2011].

Angoff, W.H. 1971. Scales, norms, and equivalent scores. In R.L. Thorndike (Ed.), *Educational Measurement* (2nd ed., pp. 508-600). Washington DC: American Council on Education.

Angoff, W.H. 1984. *Scales, Norms, and Equivalent Scores.* Princeton, NJ: Educational Testing Service.

Anstrom, K., DiCerbo, P., Ferraro, M., and Stoops Verplaetse, L. 2009, June 19. *Webinar: Academic Language in Secondary-Level Standards and Classrooms* [National Clearinghouse for English Language Acquisition]. Available: http://www.ncela.gwu.edu/webinars/event/11/ [April 2011].

Arizona Department of Education. 2007. *Arizona Technical Manual.* Office of English Language Acquisition Services. Available: http://www.ade.az.gov/oelas/AZELLA/AZELLAAZ-1TechnicalManual.pdf [October 2009].

Arkansas Department of Education. n.d. *Guide to Understanding Scores: Limited English Proficient, Arkansas.* Available: http://www.arkansased.org/lep/pdf/elda_scores_guide_051509.pdf [December 2010].

Bailey, A.L., and Heritage, H.M. 2008. *Formative Assessment for Literacy, Grades K-6: Building Reading and Academic Language Skills Across the Curriculum.* Thousand Oaks, CA: Corwin Press.

Bailey, A.I., and Kelly, K.R. 2010, July. *The Use and Validity of Home Language Surveys in State English Proficiency Assessment Systems: A Review and Issues Perspective.* White paper prepared for the U.S. Department of Education for the Evaluating the Validity of English Language Proficiency Assessments project.

Bauman, J., Boals, T., Cranley, E., Gottlieb, M., and Kenyon, D. 2007. Assessing Comprehension and Communication in English State to State for English Language Learners (ACCESS for ELLs®). In J. Abedi (Ed.), *English Language Proficiency Assessment in the Nation: Current Status and Future Practice* (pp. 81-91). Davis: University of California.

California Department of Education. 2008a. *Data Elements, Templates, and Data Files for the School Accountability Report Card (SARC) for 2008-2009.* Available: http://www.cde.ca.gov/ta/ac/sa/sarc0809.asp [January 2011].

California Department of Education. 2008b. *CELDT Released Test Questions.* Available: http://www.cde.ca.gov/ta/tg/el/documents/celdtrtqs0408.pdf [December 2010].

California Department of Education. 2008c. *Technical Report for the California Language Development Test, 2007-2008 Edition, Form G. California Language Development Test.* Available: http://www.cde.ca.gov/ta/tg/el/documents/techrpt0708.pdf [December 2010].

California Department of Education. 2009a. *Communicating 2009-2010 Individual Results with Parents and Guardians.* Available: http://www.cde.ca.gov/ta/tg/el/documents/celdt09astpkt2rev.pdf [December 2010].

California Department of Education. 2009b. *Questions and Answers about the California English Language Development Test.* Available: http://www.cde.ca.gov/ta/tg/el/documents/celdt09qa.pdf [December 2010].

California Department of Education. 2009c. *Technical Report for the California Language Development Test, 2008-2009 Edition.* Available: http://www.cde.ca.gov/ta/tg/el/documents/techrpt0809.pdf [December 2010].

California Department of Education. 2010a. *Student National Origin Report Submission in 2009-2010.* Available: http://www.education.ca.gov/sp/el/t3/snorpinltr09.asp?print=yes [December 2010].

California Department of Education. 2010b. *Test Results Interpretation Guide for Kindergarten and Grade One, 2010-2011 Edition.* Available: http://www.celdt.org/documents/2010-11/CELDT_TRIG_2010-11_English_K-1.pdf [December 2010].

Carlton, S. 2009. *Fairness Review of Assessments.* Princeton, NJ: Educational Testing Service.

Cizek, G.J. (Ed). 2001. *Setting Performance Standards: Concepts, Methods, and Perspectives.* Mahwah, NJ: Lawrence Erlbaum.

Colorado Department of Education. 2007a. *Language, Culture, and Equity.* Available: http://www.cde.state.co.us/cde_english/download/CELA/CELA_PostTest%20Training%20(2).pdf [December 2010].

Colorado Department of Education. 2007b. *Language, Culture, and Equity. CELA Program Administration.* Available: http://www.cde.state.co.us/cde_english/download/CELA/Brochures/CELAbrochure07.pdf [December 2010].

Cook, H.G., Hicks, E., Lee, S., and Freshwater, R. 2009, July. *Methods for Establishing English Language Proficiency Using State Content and Language Proficiency Assessments.* Unpublished white paper, Wisconsin Center for Education Research, University of Wisconsin-Madison.

CTB/McGraw-Hill. 2006. Technical Manual for LAS Links. Monterey, CA: Author.

CTB/McGraw-Hill. 2009a. *LAS Links Details.* Available: http://www2.ctb.com/sites/laslinks/details/details.shtml [December 2010].

CTB/McGraw-Hill. 2009b. *LAS Links Form B.* Missouri Index of PDFs. Available: http://www.mo-ell.org/pdfs/macii/LASLinksTraining.pdf.

CTB/McGraw-Hill. 2009c. *LAS Links K-12 Assessments.* Available: http://www.ctb.com/products/product_detail.jsp?FOLDER%3C%3Efolder_id=1408474395292398&bmUID=1259010098152.

CTB/McGraw-Hill. 2009d. *LAS Links Placement Test.* Available: http://www.ctb.com/products/product_summary.jsp?FOLDER%3C%3Efolderid=1408474395292913&bmUID=1259010100589.

De Ávila, E. 1997. *Setting Expected Gains for Non- and Limited-English Proficient Students.* NCELA Resource Collection Series No. 8. Washington, DC: National Clearinghouse for English Language Acquisition.

Del Vecchio, A.N., and Guerrero, M. 1995. *Handbook of English Language Proficiency Tests.* Washington, DC: National Clearinghouse for Bilingual Education.

EdSource. 2008. *English Learners in California: What the Numbers Say.* Available: http://www.edsource.org/pub_ELvitalstats3-08.html [December 2010].

Educational Testing Service. 2005. *CELLA Technical Summary Report.* Available: http://www.accountabilityworks.org/photos/CELLA_Technical_Summary_Report.pdf [December 2010].

Ercikan, K., Schwarz, R.D., Julian, M.W., Burkett, G., Weber, M.M., and Link, V. 1998. Calibration and scoring of tests with multiple-choice and constructed-response item types. *Journal of Educational Measurement, 35*(2), 137-154.

Federal Register. 2008. *Title III of the Elementary and Secondary Education Act of 1965 as Amended by the No Child Left Behind Act of 2001.* Doc E8-24702, vol. 73, no. 202, 61828-61844. Available: http://www.federalregister.gov/articles/2008/10/17/E8-24702/title-iii-of-the-elementary-and-secondary-education-act-of-1965-esea-as-amended-by-the-no-child-left [April 2011].

Finkleman, M. n.d. *Decision Accuracy and Consistency.* Available: http://www.measuredprogress.org/resources/psychometrics/learningtools/intermediate/materials/DecisionAccuracy.pdf.

Fitzpatrick, A.R., Julian, M.W., Hauck, M.C., and Dawber, T.E. 2006. *The Dimensionality of Two NCLB Tests Designed to Assess Students' English Language Proficiency.* Paper presented at the annual meeting of the American Educational Research Association, San Francisco, CA.

Florida Department of Education. n.d. *CELLA Information. Florida-Comprehensive English Language Learning Assessment (CELLA).* Available: http://www.fldoe.org/aala/pdf/2008CELLAInfoFlyerENG.pdf [December 2010].

Florida Department of Education. 2006. *CELLA Fact Sheet. Florida-Comprehensive English Language Learning Assessment (CELLA).* Available: http://www.fldoe.org/aala/pdf/cellainfosheet.pdf [December 2010].

Florida Department of Education. 2008. *CELLA Score Reports. Florida-Comprehensive English Language Learning Assessment (CELLA).* Available: http://www.fldoe.org/aala/pdf/2008%20FL%20CELLA%20Score%20Report%20Interpretation_061108FINAL.

Florida Department of Education. 2009. *CELLA Interpretive Guide-English. Florida-Comprehensive English Language Learning Assessment (CELLA).* Available: http://www.fldoe.org/aala/pdf/CELLA-Interpretive-Guide.pdf [December 2010].

Francis, D.J., and Rivera, M.O. 2007. Principles underlying English language proficiency tests and academic accountability for ELLs. In J. Abedi (Ed.), *English Language Proficiency Assessment in the Nation: Current Status and Future Practice* (pp. 13-31). Davis: University of California.

Garrett, D., Williams, A., and Petroni, R. 2004. *Monitoring, Experience and Performance of CATI Interviewers in the American Community Survey.* In Joint Statistical Meeting Proceedings. Available: http://www.amstat.org/sections/srms/proceedings/y2004/files/Jsm2004-000521.pdf [December 2010].

Garson, D. 2009. *Reliability Analysis.* North Carolina State University. Available: http://faculty.chass.ncsu.edu/garson/PA765/reliab.htm [December 2010].

Haertel, E.H., and Linn, R.L. 1996. Comparability. In G.W. Phillips (Ed.), *Technical Issues in Large-Scale Performance Assessment* (pp. 59-78). Report No. NCES 96-802. Washington, DC: Office of Educational Research and Improvement, U. S. Department of Education.

Hakuta, K., Butler, Y., and Witt, D. 2000. *How Long Does It Take English Learners to Attain Proficiency?* Linguistic Minority Research Institute Policy Report 2000-1. Santa Barbara: University of California.

Harcourt. 2007. *AZELLA Technical Manual, Form AZ-1.* Available: http://www.ade.az.gov/oelas/AZELLA/AZELLAAZ-1TechnicalManual.pdf [April 2011].

Harris, P., Bhaskar, R., Shook-Finucane, C., and Ericson, L. 2007. *Evaluation Report Covering Place of Birth, U.S. Citizenship Status, and Year of Arrival: Final Report.* Available: http://www.census.gov/acs/www/AdvMeth/content_test/P1_Citizenship.pdf [December 2010].

Holland, P.W., and Dorans, N.J. 2006. Linking and equating. In R.L. Brennan (Ed.), *Educational Measurement* (4th ed., pp. 187-220). Westport, CT: American Council on Education/Praeger.

Illinois State Board of Education. 2009. *Immigrant Education Program.* Available: http://www.isbe.net/bilingual/pdfs/immigrant_program_app.pdf [December 2010].

Impara, J.C., and Plake, B.S. 1997. Standard setting: An alternative approach. *Journal of Educational Measurement, 34*, 355-368.

Jaeger, R.M. 1989. Certification of student competence. In R.L. Linn (Ed.), *Educational Measurement* (3rd ed., pp. 485-511).Washington, DC: American Council on Education.

Johnson, E.G., and Owen, E. 1998. Types of linkage. Chapter 3 in *Linking the National Assessment of Educational Progress (NAEP) and the Third International Mathematics and Science Study (TIMSS): A Technical Report.* Washington, DC: National Center for Education Statistics. Available: http://nces.ed.gov/pubs98/98499.pdf [December 2010].

Kane, M.T. 1995. So much remains the same: Conception and status of validation in setting standards. In G.J. Cizek (Ed.), *Setting Performance Standards: Concepts, Methods and Perspectives* (pp. 53-88). Mahwah, NJ: Lawrence Erlbaum.

Kane, M.T. 2002. Conducting standard-setting studies based on standards of practice. *The Bar Examiner, November.*

Kenyon, D.M. 2006a. *The Bridge Study Between Tests of English Proficiency and ACCESS for ELLS, Part IIA: IPT Results.* Available: http://www.wida.us/assessment/ACCESS/TechReports/Bridge%20Study%20Report%20Part%202A%20-%20IPT%20Results.pdf [December 2010].

Kenyon, D.M. 2006b. *The Bridge Study Between Tests of English Proficiency and ACCESS for ELLS, Part IIB: LAS Results.* Available: http://www.wida.us/assessment/ACCESS/TechReports/Bridge%20Study%20Report%20Part%202B%20-%20LAS%20Results.pdf [December 2010].

Kenyon, D.M. 2006c. *The Bridge Study Between Tests of English Proficiency and ACCESS for ELLS, Part IID: MAC II Results.* Available: http://www.wida.us/assessment/ACCESS/TechReports/Bridge%20Study%20Report%20Part%202D%20-%20MAC%20II%20Results.pdf [December 2010].

Kenyon, D.M. 2006d. *The Bridge Study Between Tests of English Proficiency and ACCESS for ELLS, Part IIC: LPTS Results.* Available: http://www.wida.us/assessment/ACCESS/TechReports/Bridge%20Study%20Report%20Part%202C%20-%20LPTS%20Results.pdf [December 2010].

Kim, J., and Herman, J.L. 2009. *A Three-State Study of English Learner Progress.* Los Angeles: National Center for Research on Evaluation, Standards, and Student Testing, University of California.

Kingston, N.M., Kahl, S.R., Sweeney, K.P. and Bay, L. 2001. Setting performance standards using the body of work method. In G.J. Cizek (Ed.), *Setting Performance Standards: Concepts, Methods, and Perspectives.* Mahwah, NJ: Lawrence Erlbaum.

Kiplinger, V.L. 1996. *Investigation of Factors Affecting Mathematics Achievement in the Eighth Grade: Academic Performance in Hawaii's Public Schools.* Unpublished dissertation, University of Colorado at Boulder.

Kolen, M.J., and Brennan, R.L. 2004. *Test Equating, Scaling, and Linking: Methods and Practices* (2nd ed.). New York: Springer.

Kominski, R. 1989. *How Good Is "How Well"? An Examination of the Census English Speaking Ability Question.* Paper presented at the Annual Meeting of the American Statistical Association, U.S. Census Bureau, Washington, DC.

Kubiszyn, T., and Borich, G. 2007. *Educational Testing and Measurement* (8th ed.). Hoboken, NJ: John Wiley and Sons.

Lara, J., Ferrara, S., Calliope, M., Sewell, D., Winter, P., Kopriva, R., et al. 2007. The English Language Development Assessment (ELDA). In J. Abedi (Ed.), *English Language Proficiency Assessment in the Nation: Current Status and Future Practice* (pp. 47-60). Davis: University of California.

Lewis, D.M., Mitzel, H.C., and Green, D.R. 1996. Standard setting: A bookmark approach. In D.R. Green (Chair), *IRT-Based Procedures Using Behavioral Anchoring* Symposium conducted at the Council of Chief School Officers National Conference on Large Scale Assessment, Phoenix, AZ.

Linn, R.L. 1993. Linking results of distinct assessments. *Applied Measurement in Education, 6*, 83-102.

Livingston, S.A., and Zieky, M.J. 1982. *Passing Scores: A Manual for Setting Standards of Performance on Educational and Occupational Tests.* Princeton, NJ: Educational Testing Service.

Loomis, S.C. 2001. *Judging Evidence of the Validity of the National Assessment of Educational Progress Achievement Levels.* Paper presented at the Annual Meeting of the Council of Chief State Schools Officers, Houston, TX.

MacGregor, D., Kenyon, D., Gibson, S., and Evans, E. 2009. *Development and Field Test of Kindergarten ACCESS for ELLs®*. Madison, WI: WIDA Consortium.

Mathews, G. 2007. Developing the Mountain West Assessment. In J. Abedi (Ed.), *English Language Proficiency Assessment in the Nation: Current Status and Future Practice* (pp. 33-45). Davis: University of California.

McArthur, E.K. 1991. *Characteristics of Persons in the United States by Language Spoken: 1979 and 1989.* Unpublished paper, National Center for Education Statistics, Washington, DC.

Mislevy, R.J. 1992. *Linking Educational Assessments: Concepts, Issues, Methods and Prospects.* Princeton, NJ: Educational Testing Service.

Mitzel, H.C., Lewis, D.M., Patz, R.J. and Green, D.R. 2001. The bookmark procedure: Psychological perspectives. In G.J. Cizek (Ed), *Setting Performance Standards: Concepts, Methods, and Perspectives* (pp. 249-281). Mahwah, NJ: Lawrence Erlbaum.

Musick, M.D. 2000. *Can we talk?...About how to make education standards high enough in your state.* Atlanta, GA: Southern Regional Educational Board. Retrieved May 3, 2010 from http://www.sreb.org/main/highschools/accountability/settingstandardshigh.asp.

National Research Council. 1995. *Modernizing the U.S. Census.* Panel on Census Requirements in the Year 2000 and Beyond, B. Edmonston and C. Schultze (Eds.). Committee on National Statistics. Commission on Behavioral and Social Sciences and Education. Washington, DC: National Academy Press.

National Research Council. 1999a. *Embedding Questions: The Pursuit of a Common Measure in Uncommon Tests.* D.M. Koretz, M.W. Bertenthal, and B.F. Green (Eds.). Committee on Embedding Common Test Items in State and District Assessments. Commission on Behavioral and Social Sciences and Education. Washington, DC: National Academy Press.

National Research Council. 1999b. *Uncommon Measures: Equivalence and Linkage Among Educational Tests.* M.J. Feuer, P.W. Holland, B.F. Green, M.W. Bertenthal, and F. Cadelle Hemphill (Eds.). Committee on Equivalency and Linkage of Educational Tests. Board on Testing and Assessment. Commission on Behavioral and Social Sciences and Education. Washington, DC: National Academy Press.

National Research Council. 2001. *Choosing the Right Formula: Initial Report.* Panel on Formula Allocations. T.B. Jabine, T.A. Louis, and A.L. Schirm (Eds.). Committee on National Statistics. Division of Behavioral and Social Sciences and Education. Washington, DC: National Academy Press.

National Research Council. 2004. *The 2000 Census: Counting Under Adversity.* Panel to Review the 2000 Census. C.F. Citro, D.L. Cork, and J.L. Norwood (Eds.). Committee on National Statistics. Division of Behavioral and Social Sciences and Education. Washington, DC: The National Academies Press.

National Research Council. 2005. *Measuring Literacy: Performance Levels for Adults.* Committee on Performance Levels for Adult Literacy. R.M. Hauser, C.F. Edley, Jr., J.A Koenig, and S.W. Elliott (Eds.). Board on Testing and Assessment, Center for Education. Division of Behavioral and Social Sciences and Education. Washington, DC: The National Academies Press.

National Research Council. 2007. *Using the American Community Survey—Benefits and Challenges*, 2007. Panel on the Functionality and Usability of Data from the American Community Survey. C.F. Citro and G. Kalton (Eds.). Committee on National Statistics. Division of Behavioral and Social Sciences and Education. Washington, DC: The National Academies Press.

New Jersey Department of Education. 2010. *Immigrant Student Count.* Available: http://homeroom.state.nj.us/isge/UserManual2010.htm [December 2010}.

New York State Department of Education. 2006. *New York State English as a Second Language Achievement Test, Technical Manual.* Available: http://www.p12.nysed.gov/osa/reports/nyseslat-techrep-06.pdf [December 2010].

New York State Department of Education. 2009a. *New York State English as a Second Language Achievement Test.* Available: http://www.p12.nysed.gov/osa/nyseslat/sam-cr09.pdf [December 2010].

New York State Department of Education. 2009b. *Performance Indicators by Grade Level. The Teaching of Language Arts to Limited English Proficient /English Language Learners: Learning Standards for English as a Second Language.* Available: http://www.p12.nysed.gov/biling/resource/ESL/standards.html [December 2010].

New York State Department of Education. 2009c. *New York State English as a Second Language Achievement Test School, Administrator's Manual.* Available: http://www.emsc.nysed.gov/osa/nyseslat/sam-cr09.pdf [December 2010].

New York State Department of Education. 2011. *NYSESLAT Parent's Guide.* Available: http://www.p12.nysed.gov/osa/nyseslat/ [February 2011).

Ohio Department of Education. 2009. *Ohio Test of English Language Acquisition.* Ohio Department of Learning. Available: http://education.ohio.gov/GD/Templates/Pages/ODE/ODEDetail.aspx?page=3&TopicRelationID=500&ContentID=8402&Content=69206 [December 2010].

Organisation for Economic Co-operation and Development. 2003. *Quality Framework and Guidelines for OECD Statistical Activities.* Version 2003/1. Available: http://www.oecd.org/dataoecd/26/42/21688835.pdf [December 2010].

Parker, C.E., Louie, J., and O'Dwyer, L. 2009. *New Measures of English Language Proficiency and Their Relationship to Performance on Large-Scale Content Assessments* (Issues & Answers Report, REL 2009-No. 066). Washington, DC: U.S. Department of Education, Institute of Education Sciences, National Center for Education Evaluation and Regional Assistance, Regional Educational Laboratory Northeast and Islands. Available: http://ies.ed.gov/ncee/edlabs [December 2010].

Passel, J.S., Hook, J.V., and Bean, F.D. 2006. *Estimates by Migrant Status: Narrative Profile with Adjoining Tables of Unauthorized Migrants and Other Immigrants, Based on Census 2000: Characteristics and Methods.* U.S. Bureau of the Census and Sabre Systems Statistical and Demographic Analyses, Immigration Studies White Papers. Available: http://www.sabresystems.com/sd_whitepapers_immigration.asp [December 2010].

Porta, S., and Vega, J. 2007. Appendix A: Overview of existing English Language proficiency tests. In J. Abedi (Ed.), *English Language Proficiency Assessment in the Nation: Current Status and Future Practice* (pp. 133-189). Davis, CA: University of California, School of Education. Available: http://education.ucdavis.edu/research/ELP_Report.pdf [December 2010].

Professional Testing Incorporated. 2006. *Test Reliability.* Available: http://www.proftesting.com/test_topics/pdfs/test_quality_reliability.pdf [December 2010].

Reamer, A. 2010. *Surveying for Dollars: The Role of the American Community Survey in the Geographic Distribution of Federal Funds.* Metropolitan Policy Program. Washington, DC: The Brookings Institution.

Rebarber, T., Rybinski, P., Hauck, M., Scarcella, R., Buteux, A., Wang, J., et al. 2007. Designing the Comprehensive English Language Learner Assessment (CELLA) for the benefit of users. In J. Abedi (Ed.), *English Language Proficiency Assessment in the Nation: Current Status and Future Practice* (pp. 63-79). Davis: University of California.

Reckase, M.D. 2000. *The Evolution of the NAEP Achievement Levels Setting Process: A Summary of the Research and Development Efforts Conducted by ACT.* Iowa City, IA: Act, Inc.

Redstone, I., and Massey, D. 2003. *Coming to Stay: An Analysis of the U.S. Census Question on Year of Arrival.* Paper presented at the Population Association of America Annual Meeting, May 1-3, Minneapolis, MN.

Regan, A., and Lesaux, N. 2006. Federal, state, and district level English language learner program entry and exit requirements: Effects on the education of language minority learners. (20). Available: http://www.doaj.org/doaj?func=abstract&id=683508&recNo=20&toc=1&uiLanguage=en [April 2011].

Rossell, C.H. 2000. *Different Questions, Different Answers: A Critique of the Hakuta, Butler, and Witt Report—How Long Does It Take English Learners to Attain Proficiency? Read Perspectives.* Washington, DC: Education Resources Information Center.

Scarcella, R. 2008. *Conversational Analysis in L2 Acquisition and Teaching.* Cambridge, England: Cambridge University Press.

Schmidley, D., and Robinson, J.G. 2003. *Measuring the Foreign-Born Population in the United States with the Current Population Survey: 1994-2002.* Population Division working paper number 73. Washington, DC: U.S. Census Bureau.

Shin, H.B. 2008. *Comparison of the Estimates on Language Use and English Speaking Ability for the ACS, the C2SS, and Census 2000.* Washington, DC: U.S. Census Bureau. Available: http://www.census.gov/acs/www/Downloads/library/2008/Language_Comparison_Report_2008-03.pdf [December 2010].

Short, D.J., and Fitzsimmons, S. 2007 A report to Carnegie Corporation of New York. Washington, DC: Alliance for Excellent Education. Available: http://www.all4ed.org/files/DoubleWork.pdf [April 2011].

Siegel, P., Martin, E., and Bruno, R. 2001. *Language Use and Linguistic Isolation: Historical Data and Methodological Issues.* Prepared for the session on Language Differences and Linguistic Isolation at the FCSM Statistical Policy Seminar, November 8-9, 2000, Bethesda, MD.

Statistics Canada. 2009. *Statistics Canada Quality Guidelines* (5th ed.), Catalogue No. 12-539-X. Available: http://www.statcan.gc.ca/pub/12-539-x/12-539-x2009001-eng.pdf [December 2010].

Stephenson, A. 2003. *Establishing Performance Levels for the Standard English Language Proficiency Test.* San Antonio, TX: Harcourt Assessment.

Texas Education Agency. n.d. *TELPAS Proficiency Level Descriptors. TELPAS English Language Proficiency Assessment System.* Available: http://ritter.tea.state.tx.us/student.assessment/ELL/TELPAS-PLDs.pdf [December 2010].

Texas Education Agency. 2002. *State Accountability Data Tables, Base Indicators.* Austin: Texas Department of Education.

Texas Education Agency. 2008a. *Appendix C. Technical Digest 2007-2008.* Available: http://ritter.tea.state.tx.us/student.assessment/resources/techdigest/2008/appendix_c.pdf [December 2010].

Texas Education Agency. 2008b. *Chapter 16: Reliability. Technical Digest 2007-2008.* Available: http://ritter.tea.state.tx.us/student.assessment/resources/techdigest/2008/chapter_16.pdf [December 2010].

Texas Education Agency. 2008c. *Chapter 17: Validity. Technical Digest 2007-2008.* Available: http://ritter.tea.state.tx.us/student.assessment/resources/techdigest/2008/chapter_17.pdf [December 2010].

Texas Education Agency. 2008d. *TELPAS Reading Information Booklet. TELPAS English Language Proficiency Assessment System.* Available: http://ritter.tea.state.tx.us/student.assessment/admin/rpte/TP08_InfoBook.pdf [December 2010].

Texas Education Agency. 2009a. *Manual for Raters and Administrators. TELPAS English Language Proficiency Assessment System.* Available: http://ritter.tea.state.tx.us/student.assessment/resources/guides/test_admin/2009/TELPAS_Manual_2009.pdf [December 2010].

Texas Education Agency. 2009b. *Program Overview.* Available: http://ritter.tea.state.tx.us/student.assessment/resources/guides/interpretive/2009_Interpretive_Guide_TELPAS.pdf [December 2010].

Texas Education Agency. 2009c. *Technical Digest 2008-2009.* Available: http://www.tea.state.tx.us/index3.aspx?id=2147484418&menu_id=793 [December 2010].

U.S. Census Bureau. 2007. Weighting and estimation. In *Design and Methodology Report.* Available: http://www.census.gov/acs/www/Downloads/survey_methodology/acs_design_methodology_ch11.pdf [December 2010].

U.S. Census Bureau. 2010. *U.S. Census Bureau Statistical Quality Standards.* Available: http://www.census.gov/quality/standards/Quality_Standards_June2010.pdf [December 2010].

U.S. Department of Commerce. 1978. *Report on Statistics for the Allocation of Funds.* Federal Committee on Statistical Methodology, Statistical Policy Working Paper No. 1. Washington, DC: U.S. Government Printing Office. Available: http://www.fcsm.gov/working-papers/spwp1.html [December 2010].

U.S. Department of Education. 1987. *Numbers of Limited English Proficient Children: National, State, and Language-Specific Estimates.* Washington, DC: Office of Planning, Budget and Evaluation.

U.S. Department of Education. 2005a. *Biennial Evaluation Report to Congress on the Implementation of Title III, Part A of the ESEA.* Office of English Language Acquisition, Language Enhancement, and Academic Achievement for Limited English Proficient Students. Washington, DC: Author.

U.S. Department of Education. 2005b. *Letter to Title III State Directors.* Office of English Language Acquisition, Language Enhancement, and Academic Achievement for Limited English Proficient Students. Washington, DC: Author. Available: http://wvconnections.k12.wv.us/documents/TimelineforELPAssessment.doc [December 2010].

U.S. Department of Education. 2008a. *Biennial Report to Congress on the Implementation of the Title III State Formula Grant Program, School Years 2004-2006.* Office of English Language Acquisition, Language Enhancement, and Academic Achievement for Limited English Proficient Students. Washington, DC: Author.

U.S. Department of Education. 2008b. *Notice of Final Interpretations. Title III of the Elementary and Secondary Education Act of 1965, as Amended by the No Child Left Behind Act of 2001.* Office of English Language Acquisition. Washington, DC: Author.

U.S. Department of Education. 2010a. *EDFacts Workbook SY 2009-2010 (Version 6.0).* Available: http://www2.ed.gov/about/inits/ed/edfacts/eden/09-10-workbook-6-0.pdf [December 2010].

U.S. Department of Education. 2010b. *Title III Policy. State of the States. ESEA Evaluation Brief: The English Language Acquisition, Language Enhancement, and Academic Achievement Act.* Washington, DC: Author.

U.S. Government Accountability Office. 2006a. *Assistance from Education Could Help States Better Measure Progress of Students with Limited English Proficiency.* GAO 06-815. Washington, DC: Author.

U.S. Government Accountability Office. 2006b. *Education's Data Improvement Efforts Could Strengthen the Basis for Distributing Title III Funds.* GAO-07-140. Washington, DC: Author.

Washington State Department of Education. 2008a. *Language Proficiency Levels. ELD Content Standards—State of Washington.* Available: http://www.k12.wa.us/MigrantBilingual/ELD.aspx [December 2010].

Washington State Department of Education. 2008b. *Washington Proficiency Levels.* Washington Assessment and Testing Web. Available: htttp://www.k12.wa.us/assessment/WLPTII/pubdocs/2008/2008EnglishTranslationLetter.pdf.

Washington State Department of Education. 2008c. *WLPT-II Placement Test Training.* Available: http://www.k12.wa.us/assessment/WLPTII/pubdocs/2008/WLPT-IIPlacementTestOverview.pdf.

Washington State Department of Education. 2009a. *WLPT-II Cut Scores.* Available: http://www.k12.wa.us/assessment/WLPTII/pubdocs/2009/2009WLPT-IICutScaleScores.pdf.

Washington State Department of Education. 2009b. *WLPT-II Technical Report.* Available: http://www.k12.wa.us/assessment/pubdocs/WLPT-II2009TechnicalReport.pdf [December 2010].

WIDA Consortium. 2006a. *Annual Technical Report No. 1-Volume 1 of 3: Description, Validity, and Student Results (2004-2005).* Technical Reports and Technical Advisory Committee (TAC). Available: http://www.wida.us/assessment/ACCESS/TechReports/1%20Tech%20Rpt%201%20Series%20100.pdf [December 2010].

WIDA Consortium. 2006b. *Technical Report No. 1-Development and Field Test of ACCESS for ELLs.* Madison, WI: Author.

WIDA Consortium. 2006c. *The Bridge Study between Tests of English Language Proficiency and ACCESS for ELLs. Part II A: IPT Results Technical Report No. 1.* Available: http://www.wida.us/assessment/ACCESS/TechReports/Bridge%20Study%20Report%20Part%202A%20-%20IPT%20Results.pdf [December 2010].

WIDA Consortium. 2006d. *The Bridge Study between Tests of English Language Proficiency and ACCESS for ELLs. Part II B: LAS Results Technical Report No. 2.* Available: http://www.wida.us/assessment/ACCESS/TechReports/Bridge%20Study%20Report%20Part%202B%20-%20LAS%20Results.pdf [December 2010].

WIDA Consortium. 2006e. *The Bridge Study between Tests of English Language Proficiency and ACCESS for ELLs. Part II C: LPTS Results Technical Report No. 2.* Available: http://www.wida.us/assessment/ACCESS/TechReports/Bridge%20Study%20Report%20Part%202C%20-%20LPTS%20Results.pdf [December 2010].

WIDA Consortium. 2006f. *The Bridge Study between Tests of English Language Proficiency and ACCESS for ELLs. Part II D: MAC II Results Technical Report No. 2.* Available: http://www.wida.us/assessment/ACCESS/TechReports/Bridge%20Study%20Report%20Part%202D%20-%20MAC%20II%20Results.pdf [December 2010].

WIDA Consortium. 2007. *Understanding the WIDA ELP Standards.* Available: http://www.wida.us/standards/Resource.

WIDA Consortium. 2009a. *ACCESS for ELLs Teacher Report. WIDA Site-North Carolina.* Madison, WI: Author.

WIDA Consortium. 2009b. *Annual Technical Report No. 4-Volume 1 of 3: Description, Validity, and Student Results (2007-2008).* Technical Reports and Technical Advisory Committee (TAC). Available: http://www.wida.us/assessment/ACCESS/TechReports/1%20Tech%20Rpt%204%20Series%20103.pdf [December 2010].

WIDA Consortium. 2009c. *Interpretive Guide for Score Reports. ACCESS for ELLs Score Reports.* Available: http://www.wida.us/assessment/ACCESS/ScoreReports/ACCESS_Interpretive_Guide09.pdf [December 2010].

WIDA Consortium. 2009d. WIDA tally: Hawaii joins testing consortium. *WIDA News*, July 8. Available: http://www.wida.us/index.aspx [December 2010].

WIDA Consortium. 2009e. WIDA tally: New Mexico joins consortium. *WIDA News*, November 2. Available: http://www.wida.us/index.aspx [December 2010].

WIDA Consortium. 2009f. *White Paper on Methods of Establishing English Language Proficiency using State Content and Language Proficiency Assessments.* Madison, WI: Author.

Wolf, M., Kao, J., Griffin, N., Herman, J., Bachman, P., Chang, S., and Farnsworth, T. 2008. *Issues in Assessing English Language Learners: English Language Proficiency Measures and Accommodation Uses: Literature Review.* Available: http://www.cse.ucla.edu/products/rsearch.asp [December 2010].

Zehler, A.M., Hopstock, P.J., Fleischman, H.L., and Greniuk, C. 1994. *An Examination of Assessment of Limited English Proficient Students.* Arlington, VA: Development Associates, Special Issues Analysis Center.

Zeiky, M. 2001. So much has changed: How the setting of cutscores has evolved since the 1980s. In G. Cizek (Ed.), *In Setting Performance Standards: Concepts, Methods and Perspectives* (pp. 19-51). Mahwah, NJ: Lawrence Erlbaum.

Appendix A

Review of English Language Proficiency Tests

As part of the panel's work, we identified eight English language proficiency (ELP) tests to review in detail (see Chapter 3). These eight tests are used by 40 states and are administered to approximately 75 percent of the English language learner (ELL) students in the country. The tests that we reviewed are listed in Table A-1 along with the states that used each of them during the 2009-2010 school year.

Our review is based on several sources of information. First, we reviewed the technical manuals available for each test. Second, we consulted two recent reports that summarized technical information about the tests: Abedi (2007) provides detailed information about each of the consortium-developed ELP tests (as explained in Chapter 3) and brief descriptions of all of the tests used by the states during the 2006-2007 school year; Wolf et al. (2008) provide a summary of technical information available for 13 ELP tests available as of 2007. Third, representatives from four testing programs—Assessing Comprehension and Communication in English State-to-State (ACCESS), the English Language Development Assessment (ELDA), Language Assessment Scales Links K-12 (LAS-Links), and the Stanford English Language Proficiency Test (SELP)—met with the panel at our second meeting to discuss their tests. This appendix summarizes the information we obtained from these sources.

ASSESSING COMPREHENSION AND COMMUNICATION STATE TO STATE FOR ELL STUDENTS

ACCESS was developed by the World-Class Instructional Design and Assessment (WIDA) Consortium. It began as a partnership of three states—Arkansas, Delaware, and Wisconsin—with technical support through the Center for Applied

TABLE A-1 English Language Proficiency Tests Reviewed and the States That Use Them

Test	States Using the Test During the 2009-2010 School Year
ACCESS	Alabama, Delaware, DC, Georgia, Hawaii, **Illinois**, Kentucky, Maine, Mississippi, Missouri, New Hampshire, New Jersey, New Mexico, **North Carolina**, North Dakota, Oklahoma, Pennsylvania, Rhode Island, South Dakota, Vermont, **Virginia**, Wisconsin, Wyoming
CELDT	**California**
CELLA	**Florida**
ELDA	Arkansas, Iowa, Louisiana, Nebraska, South Carolina, Tennessee, West Virginia
LAS Links*	**Colorado**, Connecticut, Indiana, Maryland
NYSESLAT	**New York**
SELP[a]	**Arizona**, **Washington**
TELPAS	**Texas**
Total Tests, 8	Total states, 40

NOTE: States in bold are those with high numbers of ELL students.
[a]Test is customized for each state so that it measures the state's content standards.

Linguistics (CAL), the University of Wisconsin system, and the University of Illinois at Urbana-Champaign. Shortly after grant funding was awarded, seven other states joined the consortium (Alabama, District of Columbia, Illinois, Maine, New Hampshire, Rhode Island, and Vermont). Field-testing was done in 2004, and by spring 2005, the test was operational in three states (Alabama, Maine, and Vermont). By spring 2006, 12 states were using the assessment. At this point, development efforts were transferred from the Wisconsin Department of Public Instruction to the University of Wisconsin-Madison's Wisconsin Center for Education Research (WCER) (Bauman et al., 2007, pp. 81, 82). In the 2010-2011 testing cycle, ACCESS will be operational in 24 states. Development work on ACCESS is on-going, and approximately one-third of the test is refreshed every year.[1]

Content Standards

The ELP content standards for ACCESS were developed jointly by eight of the WIDA member states in 2003. According to Bauman and colleagues (2007), in developing the standards, the consortium wanted to ensure two essential elements: (1) a strong representation of the language of state academic standards across the

[1]Information about ACCESS is available at http://www.wida.us/assessment/access/index.aspx [December 2010].

core content areas (language arts, math, science, social studies, and the classroom setting); and (2) consensus by member states on the components of the ELP standards. As new states have joined the consortium, teams of researchers have continued the process by conducting alignment studies between the WIDA standards and a state's content standards.

Grade Bands

ACCESS reports information for five grade bands: K, 1-2, 3-5, 6-8, and 9-12. For each grade band except kindergarten, three difficulty levels of the test are available. The difficulty levels are intended to tailor the test to students' approximate proficiency range.

Item Types

ACCESS consists of both multiple-choice (the listening and reading tests) and constructed-response items (the writing and speaking tests). The speaking test is adaptive and administered one-on-one; the other tests are typically administered in a group setting. ACCESS test items are embedded in the context of a content-based theme, called a folder. A folder typically consists of a shared theme graphic followed by three or four items.

Scores Reported

ACCESS reports scores for each of the domains—reading, writing, listening, and speaking—as well as four composite scores. The overall composite score is formed by weighting reading and writing by 35 percent each and by weighting listening and speaking by 15 percent each. Reading and writing are weighted higher on the basis of the test developer's judgment about their importance for academic language proficiency. An oral language composite score is formed by equally weighting scores in listening and speaking; similarly, a literacy composite score is formed by equally weighting the scores in reading and writing. The comprehension composite score weights reading by 70 percent and listening by 30 percent (from Bauman et al., 2007, p. 90).

Performance Levels

ACCESS scores are reported using six proficiency levels: entering, beginning, developing, expanding, bridging, and reaching, defined as follows (MacGregor et al., 2009):

Entering: English language learners will process, understand, produce, or use

- pictorial or graphic representation of the language of the content areas;

- words, phrases, or chunks of language when with one-step commands, directions, use of questions, or statements with visual and graphic support.

Beginning: English language learners will process, understand, produce, or use

- general language related to the content areas;
- phrases or short sentences;
- oral or written language with phonological, syntactic, or semantic errors that often impede the meaning of the communication when presented with one, to multiple-step commands, directions, questions, or a series of statements with visual and graphic support.

Developing: English language learners will process, understand, produce, or use

- general and some specific language of the content areas;
- expanded sentences in oral interaction or written paragraphs;
- oral or written language with phonological, syntactic, or semantic errors that may impede the communication but retain much of its meaning when presented with oral or written, narrative or expository descriptions with occasional visual and graphic support.

Expanding: English language learners will process, understand, produce, or use

- specific and some technical language of the content areas;
- a variety of sentence lengths of varying linguistic complexity in oral discourse or multiple, related paragraphs;
- oral or written language with minimal phonological, syntactic, or semantic errors that do not impede the overall meaning of the communication when presented with oral or written connected discourse with occasional visual and graphic support.

Bridging: English language learners will process, understand, produce, or use

- the technical language of the content areas;
- a variety of sentence lengths of varying linguistic complexity in extended oral or written discourse, including stories, essays, or reports;
- oral or written language approaching comparability to that of English proficient peers when presented with grade level material.

Reaching: English language learners will process, understand, produce, or use

- specialized or technical language reflective of the content area at grade level;
- a variety of sentence lengths of varying linguistic complexity in extended oral or written discourse as required by the specified grade level;
- oral or written communication in English comparable to proficient English peers.

Cut scores for the levels were set using the bookmark procedure[2] for listening and reading and the body of work method[3] for writing and speaking (Bauman et al., 2007, pp. 84, 86). Following the introduction of the new pre-K cluster in 2007, an additional standard setting study for this cluster was conducted in 2008 (MacGregor et al., 2009).

The WIDA Consortium allows its member states to determine the performance level on the ACCESS they consider to be English proficient (i.e., the level that indicates the student is sufficiently proficient to be considered for reclassification as a former ELL). The levels vary by state, with some setting the proficient level at expanding, some at bridging, and some at reaching.

Reliability and Validity

Information about the technical qualities of the ACCESS assessment is provided in its technical reports, which are prepared each year; and the most recent report available to the panel was for the administrations held during the 2008-2009 school year.[4] The technical reports contain detailed information about test specifications, item and form development, item and form analysis, equating, and standard setting. They also contain results of analyses to evaluate reliability and validity, and they document efforts to evaluate fairness issues (e.g., bias review panels, analyses of differential item functioning). Reliability analyses include the standard types of analyses used for tests with multiple-choice items (i.e., estimates of internal consistency), as well as those used for open-ended items (i.e., interrater agreement, generalizabilty analyses).

A number of validity studies have been conducted to collect content-, construct-, and criterion-related evidence. Content-related validity evidence was collected by comparing *a priori* proficiency levels (the proficiency level the item was designed to target) against the item's difficulty. Expert review is also used to evaluate the extent to which items measure the intended content. Construct-related evidence consists primarily of the degree of correspondence among the subtest scores (i.e., the intercorrelations). Some evidence of criterion-related validity has been collected. One study involved comparing ACCESS scores to *a priori* ELP categorizations of students who participated in the field tests (described in Wolf et al., 2007, p. J2-75]). Another study involved comparisons of performance for students who took ACCESS and one of the older generation ELP tests, including the New IDEA Proficiency Test (New-IPT), the Language Assessment Scales (LAS), the Maculaitis Assessment of Competencies Test of English Language Proficiency (MAC II), and the Language Proficiency Test Series (LPTS).

---

[2]See Mitzel et al. (2001) for an explanation of this method.

[3]See Kingston et al. (2001) for an explanation of this method.

[4]The reports are available at http://www.wida.us/assessment/access/TechReports/index.aspx [December 2010].

CALIFORNIA ENGLISH LANGUAGE DEVELOPMENT TEST

The California English Language Development Test (CELDT) was developed and in place prior to the implementation of the No Child Left Behind Act (NCLB). In 1997, state legislation authorized the California Department of Education to develop ELP standards and a language proficiency assessment that would be used statewide, and the standards were adopted in 1999. The first version of the CELDT consisted primarily of items developed by CTB/McGraw-Hill for the Language Assessment Scales (LAS) tests, with some new items the test publisher developed specifically for the state. This version of the test was field tested in fall 2000. Data from the field test were used to select items and create the operational forms of the test, which were first administered in 2001. The CELDT has been updated yearly since 2001. Subsequent versions have replaced the LAS items with new items that are aligned with the California standards (Porta and Vega, 2007, p. 138).[5]

Content Standards

According to CELDT information, its test questions are designed to assess basic social conventions, rudimentary classroom vocabulary, and ways to express personal and safety needs. Some of the questions are designed to assess student performance at the early advanced and advanced proficiency levels and to incorporate classroom language. To this end, the questions engage academic language functions, such as explaining questions, analyzing, and summarizing.

Grade Bands

The CELDT has test versions for each of four grade bands: K-2, 3-5, 6-8, and 9-12.

Item Types

The test uses a combination of multiple-choice and constructed-response items. The reading test uses only multiple-choice items, and the speaking test uses only constructed-responses items (requiring both short and extended answers). The listening and writing tests use a combination of item types: the listening uses multiple-choice and short-answer constructed-response items; the writing uses multiple-choice, short-answer constructed-response, and extended-answer constructed-response items (California Department of Education, 2008c, 2009c).

---

[5]Information about the test is available at http://www.cde.ca.gov/ta/tg/el/ [December 2010].

Scores Reported

Scores are reported for each domain—listening, speaking, reading, and writing. Two composite scores are also reported. The comprehension score is derived from performance on the reading and listening subtests, and an overall composite score is also reported. For grades 3 through 12, the composite score is the average of the scores in all four domains. For kindergarten through grade 1, the composite score is formed by weighting listening and speaking by 45 percent each and by weighting reading and writing by 5 percent each (California Department of Education, 2009c).

Performance Levels

Five performance levels are reported for the CELDT: beginning, early intermediate, intermediate, early advanced, and advanced, as follows (California Department of Education, 2009c).

> *Beginning:* Students performing at this level of may demonstrate little or no receptive or productive English skills. They are beginning to understand a few concrete details during unmodified beginning instruction. They may be able to respond to some communication and learning demands but with many errors. Oral and written production is usually limited to disconnected words and memorized statements and questions. Frequent errors make communication difficult.
>
> *Early Intermediate:* Students performing at this level continue to develop receptive and productive English skills. They are able to identify and understand more concrete details during unmodified instruction. They may be able to respond with increasing ease to more varied communication and learning demands with a reduced number of errors. Oral and written production is usually limited to phrases and memorized statements and questions. Frequent errors still reduce communication.
>
> *Intermediate:* Students performing at this level begin to tailor the English language skills to meet communication and learning demands with increasing accuracy. They are able to identify and understand more concrete details and some major abstract concepts during unmodified instruction. They are able to respond with increasing ease to more varied communication and learning demands with a reduced number of errors. Oral and written production has usually expanded to sentences, paragraphs, and original statements and questions. Errors still complicate communication.
>
> *Early Advanced:* Students at this level begin to combine the elements of the English language in complex, cognitively demanding situations and are able to use English as a means for learning in academic domains. They are able to identify and summarize most concrete details and abstract concepts during unmodi-

fied instruction in most academic domains. Oral and written productions are characterized by more elaborate discourse and fully developed paragraphs and compositions. Errors are less frequent and rarely complicate communication.

Advanced: Students at this level communicate effectively with various audiences on a wide range of familiar and new topics to meet social and learning demands. In order to attain the English performance level of their native English speaking peers, further linguistic enhancement and refinement are still necessary. They are able to identify and summarize concrete details and abstract concepts during unmodified instruction in all academic domains. Oral and written productions reflect discourse appropriate for academic domains. Errors are infrequent and do not reduce communication.

The cut scores were set using the bookmark standard-setting procedure (Mitzel et al., 2001). The first standard setting was conducted in spring 2001, followed by a second standard setting conducted in February 2006. To be considered proficient in English on the CELDT, students need to score at the "early advanced" level or higher and have no domain scores below "intermediate."

Reliability and Validity

Information about the technical qualities of the CELDT is provided in technical reports, which are prepared each year by the contractor (CTB/McGraw-Hill); and the most recent report available to the panel was for the administrations held during the 2008-2009 school year.[6] The technical reports contain detailed information about test specifications, item and form development, item and form analysis, equating, and standard setting. They also contain results of analyses to evaluate reliability and validity, although no bias or fairness studies appear to have been done. Reliability analyses include the standard types of analyses used for tests with multiple-choice items (i.e., estimates of internal consistency) as well as those used for open-ended items (i.e., interrater agreement, generalizabilty analyses). For the current version of the test, validity studies have been conducted to collect content- and construct-related evidence. The only criterion-related evidence that has been collected was a cut-score validation study completed in 2003, which compared qualitative assessments of 600 ELL students' language ability with their CELDT scores (Wolf et al., 2008, pp. 72-79).

COMPREHENSIVE ENGLISH LANGUAGE LEARNING ASSESSMENT

The Comprehensive English Language Learning Assessment (CELLA) was developed by the English Proficiency for All Students (EPAS) consortium with the

[6]The technical reports are available at http://www.cde.ca.gov/ta/tg/el/techreport.asp) [December 2010].

assistance of the Educational Testing Service (ETS) and Accountability Works.[7] Five states initially participated in the consortium—Florida, Maryland, Michigan, Pennsylvania, and Tennessee. Field testing of the items occurred in fall 2004. At present, Florida is the only state that uses the assessment.

Content Standards

According to the developer of the assessment, Ted Rebarber (Rebarber et al., 2007), the first stage in the process was to develop a set of proficiency benchmarks, defined as a matrix of component skills, at the grade level that students are expected to attain. The benchmarks were developed based on the experience and professional judgment of researchers at Accountability Works, language researchers, and ETS test developers. The benchmarks were reviewed and approved by educators and other representatives of the five states and acted as a set of common assessment objectives (Rebarber et al., 2007, p. 68). Once the benchmarks/objectives were established, analyses were conducted to determine the extent of alignment between the benchmarks and ELP content standards of the consortium states: The aligned standards served as the basis for developing the test.

Grade Bands

The CELLA has versions of the test available for four grade bands: K-2, 3-5, 6-8, and 9-12.

Item Types

The test uses both multiple-choice and constructed-response items. The reading and listening tests consist solely of multiple-choice items. The speaking test consists solely of constructed-response items. The writing test includes a combination of both item types.

Scores Reported

The CELLA reports four scale scores: (1) a score for the reading test; (2) a score for the writing test; (3) an oral score, which is a composite of performance on the listening and speaking subtests; and (4) an overall composite score. The subtest scores are unit weighted (i.e., summed) in forming the composites. CELLA score reports for students also provide information on the raw scores (referred to as "points awarded") in several areas. These "subscores" are reported for listening/speaking and reading/writing. Score reports indicate that the raw scores can be used

[7] Information on the assessment is available at http://www.fldoe.org/aala/cella.asp [December 2010].

to evaluate students' strengths and weaknesses, but they cannot be compared across administrations.

Performance Levels

Standard setting was conducted separately for each state participating in the consortium. Florida conducted its standard setting in winter 2006 using the bookmark procedure (Mitzel et al., 2001). Four performance levels are used: beginning, low intermediate, high intermediate, and proficient (Educational Testing Service, 2005).

> *Beginning:* Beginning students speak in English and understand spoken English that is below grade level and require continuous support. Beginning students read below grade level text and require continuous support. Beginning students write below grade level and require continuous support.
>
> *Low Intermediate:* Low intermediate students speak in English and understand spoken English that is at or below grade level and require some support. Low intermediate students read at or below grade level text and require some support. Low intermediate students write at or below grade level and require some support.
>
> *High Intermediate:* High intermediate students, with minimal support, speak in English and understand spoken English that is at grade level. High intermediate students read at grade level with minimal support. High intermediate students write at grade level with minimal support.
>
> *Proficient:* Proficient students speak in English and understand spoken English at grade level in a manner similar to non-English language learners. Proficient students read at grade level text in a manner similar to non-English language learners. Proficient students write at grade level in a manner similar to non-English language learners.

Separate cut scores were set for three subscores—the oral score (listing and speaking), reading, and writing. Performance level descriptions are provided for each of these areas.

The state's policy on reclassification procedures specifies the following criteria for determining proficient performance from the composite score (Florida Department of Education, 2006):

Grade Cluster	English Proficient Composite Score
K-2	2050
3-5	2150
6-8	2200
9-12	2250

Reliability and Validity

Information about the technical qualities of the CELLA is provided in technical reports, which are prepared by the contractor (ETS).[8] The most recent report available to the panel was published in 2005. The technical report contains detailed information about test specifications, item and form development, item and form analysis, equating, and standard setting. The report also contains results of an analysis to evaluate bias and fairness (through analyses of differential item functioning). Reliability estimates are reported in the form of standard errors of measurement. No validity information is reported in the technical manual, although Porta and Vega (2007) indicate that a factor analysis study was conducted to provide construct-related validity evidence (Fitzpatrick et al., 2006, cited in Porta and Vega, 2007, p. 77).

ENGLISH LANGUAGE DEVELOPMENT ASSESSMENT

The English Language Development Assessment (ELDA) is a consortium-based test that was developed by the Council of Chief State School Officers (CCSSO) in conjunction with states in the State Collaborative on Assessment and Student Standards for Limited English Proficient students (LEP-SCASS). To develop the assessment, the consortium worked with the American Institutes for Research (AIR) and Measurement, Incorporated—with external advice from the Center for the Study of Assessment Validity and Evaluation (C-SAVE).[9] Development work occurred between fall 2002 and December 2005. Initially, 18 states were members of LEP-SCASS, and 13 states participated in the process of developing, field testing, validating, and implementing ELDA as an operational assessment (Sharon Saez, program director with the Council of Chief State School Officers, personal communication, August 2010).[10]

---

[8]The technical reports are available at http://www.accountabilityworks.org/photos/CELLA_Technical_Summary_Report.pdf [December 2010]

[9]C-SAVE was then housed at the University of Maryland and is now housed at the University of Wisconsin.

[10]Nevada was the lead state in collaboration with Georgia, Indiana, Iowa, Kentucky, Louisiana, Nebraska, New Jersey, Ohio, Oklahoma, South Carolina, Virginia, and West Virginia.

Content Standards

ELDA was designed to assess academic English, which the consortium defines as (1) language used to convey curriculum-based academic content and (2) the language of the social environment of school. Accordingly, the test items are intended to measure language skills with content drawn from language arts, math, science, technology; and social studies, although the items do not require skills in or knowledge of content in those subjects. The test items are also intended to incorporate the language required for the school environment, which covers such topics as extracurricular activities, student health, homework, classroom management, and lunch time (American Institutes for Research, 2005).

The starting point for ELDA's standards was a synthesis of the state standards in participating states and TESOL (Teachers of English to Speakers of Other Languages) standards. Of the initial 18 member states, 6 had existing ELP standards. These standards were reviewed and merged by AIR staff, and a consortium steering committee identified a common core of standards for each domain. Some states used these standards to guide the development of their own standards; others used them to review their existing standards and ensure alignment (Lara et al., 2007, p. 48).

The development process for the assessments for grades 3 through 12 was conducted separately from the development of the test for kindergarten through grade 2, although both processes followed the same steps. Both require test administrators to observe students in a variety of settings to record students' typical behaviors or responses to a set of tasks. For each test item, test administrators are provided a description for each score point (0-3) for each task that they use to determine a score. For instance, the inventory contains items such as (Lara et al., 2007, p. 57):

- Follows a two-step verbal instruction in a nonacademic setting (e.g., going to the lunchroom).
- Identifies a picture of an object with the same ending sound as "cat."
- Uses correct English words for manipulatives (content-, age-, and grade-appropriate items).

Grade Bands

ELDA has separate versions of the test for three grade bands: 3-5, 6-8, and 9-12. ELDA also provides an inventory to assess skills of students in kindergarten through grade 2, with separate inventories for kindergarten and for grades 1 and 2. The wording or the focus of the inventory tasks varies in order to be targeted either for the developmental level of kindergarteners or that of 1st and 2nd graders.

Item Types

ELDA uses a combination of multiple-choice and constructed-response items. The listening and reading subsections consist of only multiple-choice items. The writing subsection consists of both multiple-choice and short- and extended-constructed-response items. The speaking subsection consists only of oral constructed-response items.

Scores Reported

ELDA reports scores for each of the domains—reading, writing, listening, and speaking. Two composite scores are also reported: an overall score formed from scores on all four subtests and a comprehension score formed from the listening and reading subtests.

Performance Levels

ELDA uses five proficiency levels: pre-functional, beginning, intermediate, advanced, and full English proficiency (American Institutes for Research, 2005):

Pre-functional indicates that the student is beginning to

- understand short utterances;
- use gestures and simple words to communicate;
- understand simple printed material; and
- develop communicative writing skills.

Beginning indicates that the student can

- understand simple statements, directions, and questions;
- use appropriate strategies to initiate and response to simple conversation;
- understand the general message of basic reading passages; and
- compose short informative passages on familiar topics.

Intermediate indicates that the student can

- understand standard speech delivered in school and social settings;
- communicate orally with some hesitation;
- understand descriptive material within familiar contexts and some complex narratives; and
- write simple texts and short reports.

Advanced indicates that the student can

- identify the main ideas and relevant details of discussions or presentations on a wide range of topics;
- actively engage in most communicative situations familiar or unfamiliar;
- understand the context of most text in academic areas with support; and
- write multiparagraph essays, journal entries, personal/business letters, and creative texts in an organized fashion with some errors.

Full English proficiency indicates that the student can

- understand and identify the main ideas and relevant details of extended discussion or presentations on familiar and unfamiliar topics;
- produce fluent and accurate language;
- use reading strategies the same as their native English speaking peers to derive meaning from a wide range of both social and academic texts; and
- write fluently using language structures, technical vocabulary, and appropriate writing conventions with some circumlocutions.

The bookmark procedure (Mitzel et al., 2001) was used for the reading, writing, and listening domains. For the speaking test, a generalized holistic approach was used: standard setters evaluated samples of student work, placing them into one of the five categories.

The LEP-SCASS consortium established the fully English proficient level as the performance level considered to be English proficient (i.e., the level that indicates the student is sufficiently proficient to be considered for reclassification as a former ELL), although each state may set a different level as it deems appropriate.

Reliability and Validity

Information about the technical qualities of ELDA is provided in technical reports, which are available through CCSSO staff. The technical reports are prepared by the contractor (American Institutes for Research). The most recent report available to the panel was published in 2005, with a supplement published in 2006 (see American Institutes for Research, 2005). In addition, some states have prepared their own reports. The technical reports contain detailed information about test specifications, item and form development, item and form analysis, equating, and standard setting. They also contain results of analyses to evaluate reliability and validity, including studies of bias and fairness (i.e., reviews by panels, studies of differential item functioning). Reliability analyses include the standard types of analyses used for tests with multiple-choice items (i.e., estimates of internal consistency) as well as those used for open-ended items (i.e., interrater agreement, generalizabilty analyses). The technical manual does not contain results from validity analyses, but they are available in separate reports. Studies have been conducted to obtain content- and construct-related validity evidence. Several estimates of criterion-related evidence have been collected through comparison of test performance with teacher ratings, performance on the LAS, and performance on the New Idea Proficiency Test (Lara et al., 2007).

LANGUAGE ASSESSMENT SYSTEMS LINKS

The Language Assessment Systems Links (LAS Links) was developed by CTB/McGraw-Hill, which had previously developed the Language Assessment Scales

(LAS), an ELP assessment in use prior to NCLB. LAS Links was developed to comply with NCLB. The assessment is available for states to use as an "off-the-shelf" test or to be customized so that it meets a given state's needs and is aligned with its ELP content standards. Customized versions of LAS Links may be renamed by the state. For instance, Colorado's version of LAS Links is called the Colorado English Language Assessment (CELA).[11]

Content Standards

CTB/McGraw-Hill had previously developed the English Language Proficiency Assessment Standards (ELPAS), which were intended to include the primary components of the TESOL standards and the standards for English as a second language (ESL) from several states. According to the technical manual (CTB/McGraw-Hill, 2006), LAS Links tests are intended to be aligned to these standards. When a state chooses to use LAS Links, the test publisher conducts an alignment study to determine the extent to which the assessment is aligned with the state's ELP content standards. The test is then customized or augmented to ensure that the items cover the state standards (Chris Morrison, director with CTB-McGraw/Hill, personal communication, June 2009).

Grade Bands

LAS Links has versions of the test available five grade bands: K-1, 2-3, 4-5, 6-8, and 9-12.

Item Types

The assessment uses both multiple-choice and constructed-response items. The listening and reading subtests use only multiple-choice items. The speaking subtest uses only constructed-response items (both short-answer and extended-response items) and is described by the developer as "performance based." The writing subtest includes multiple-choice items, short-answer constructed-response items, and one extended constructed-response item.

Scores Reported

The assessment reports scores separately for each domain—listening, speaking, reading, writing—and provides two composite scores. The oral score summarizes performance in listening and speaking, and the comprehension score summarizes performance in reading and writing. An overall composite score is also reported.

---

[11]Information about the assessment is available at http://www.ctb.com/ctb.com/control/productFamilyViewAction?productFamilyId=454&p=products[December 2010].

Performance Levels

The assessment reports scores using five performance levels: beginning, early intermediate, intermediate, proficient, and above proficient, defined as follows:

> *Beginning:* A Level 1 student is beginning to develop receptive and productive uses of English in the school context, although comprehension may be demonstrated nonverbally or through the native language, rather than in English.
>
> *Early Intermediate:* A Level 2 student is developing the ability to communicate in English within the school context. Errors impede basic communication and comprehension. Lexical, syntactic, phonological, and discourse features of English are emerging.
>
> *Intermediate:* A Level 3 student is developing the ability to communicate effectively in English across a range of grade-level appropriate language demands in the school context. Errors interfere with communication and comprehension. Repetition and negotiation are often needed. The student exhibits a limited range of lexical, syntactic, phonological, and discourse features when addressing new and familiar topics.
>
> *Proficient:* A Level 4 student communicates effectively in English across a range of grade-level appropriate language demands in the school context, even though errors occur. The student exhibits productive and receptive control of lexical, syntactic, phonological, and discourse features when addressing new and familiar topics.
>
> *Above Proficient:* A Level 5 student communicates effectively in English, with few if any errors, across a wide range of grade-level appropriate language demands in the school context. The student commands a high degree of productive and receptive control of lexical, syntactic, phonological, and discourse features when addressing new and familiar topics.

States may rename these levels for their own purposes. Performance-level descriptions are provided for the overall composite score. According to the technical manual, the standard setting utilized a bookmark procedure (see Lewis et al., 1996) along with a policy-based review of the cut scores by a national group.

Reliability and Validity

Information about the technical qualities of the assessment is provided in technical reports, available from the publisher. The most recent report available to the panel was published in 2006 (CTB/McGraw-Hill, 2006). The technical reports contain detailed information about test specifications, item and form development, item and form analysis, equating, and standard setting. The 2006 technical report also includes reliability estimates for the multiple-choice items (estimates of internal consistency) and open-ended items (interrater correlations). The report indicates that bias was attended to during item development but does not report any results of analyses of differential item functioning. No validity studies are reported in the technical report, and such studies do not appear to have been conducted (other than a study to relate scores on the LAS Links to the earlier LAS).

NEW YORK STATE ENGLISH AS A SECOND LANGUAGE ACHIEVEMENT TEST

The New York State English as a Second Language Achievement Test (NYSESLAT) was developed by the state of New York, first in conjunction with the Educational Testing Service (ETS) and later with Harcourt Assessment Inc. (now owned by Pearson). For the 2005 test administration, items from the Harcourt ELL item bank were initially used to construct the newly developed items for the test (including items initially developed for Stanford English Language Proficiency test forms).[12]

Content Standards

According to the technical manual (New York State Department of Education, 2006, p. 10), the New York State Education Department (NYSED) developed *Learning Standards for English as a Second Language* (ESL) to meet the requirements of NCLB. Accordingly, the state's learning standards and performance indicators are derived from the domains of speaking, listening, reading, and writing and are intended to align with the state's English Language Arts standards.

Grade Bands

The NYSESLAT has versions available for five grade bands: K-1, 2-4, 5-6, 7-8, and 9-12.

Item Types

The NYSESLAT uses a combination of multiple-choice and constructed-response items. Listening and reading subtests are comprised entirely of multiple-choice items. The speaking test requires students to provide an oral constructed response. The writing test uses both multiple-choice and constructed-response items, some involving a short written response and some involving an extended response.

---

[12]Information on the tests is provided at http://www.emsc.nysed.gov/osa/nyseslat/ [December 2010].

Scores Reported

The NYSESLAT assesses skills in the domains of reading, writing, listening, and speaking. Two composite scores are reported: an oral score that combines performance on the listening and speaking tests and a written score that combines performance on the reading and writing tests.

Performance Levels

Four performance levels have been developed for the test: beginning, intermediate, advanced, and proficient. The technical manual provides descriptions only of the proficient level:

Proficient Level: Reading

- Students read English fluently and confidently and reflect upon a wide range of grade appropriate English language texts.
- Students identify and interpret relevant data, facts, and main ideas in English literary and informational texts.
- Students comprehend and analyze the author's purpose, point of view, tone, and figurative language and appropriate inferences in English.
- Students analyze experiences, ideas, information, and issues presented by others in printed English languages text, using a variety of established criteria.
- Students demonstrate inference and "beyond the text" understanding of grade-level written English language texts.
- Students interpret, predict, draw conclusions, categorize, and make connections to their own lives and other texts.

Proficient Level: Writing

- Students utilize standard written English to express ideas on a grade-appropriate level by using varied sentence structure, language patterns, and descriptive language.
- Students apply appropriate grade-level strategies to produce a variety of English language written products that demonstrate an awareness of audience, purpose, point of view, tone, and sense of voice.
- Students use written English language to acquire, interpret, apply, and transmit information.
- Students present, in written English language and from a variety of perspectives, their opinions and judgments on experiences, ideas, information, and issues.
- Students use written English for effective social communication with a wide variety of people.
- Students integrate conventions of English language grammar, usage, spelling, capitalization, and punctuation to communicate effectively about

various topics. (Minor errors in spelling grammar or punctuation do not interfere with comprehension.)
- Students self-monitor and edit their English language written work.
- Students write literary, interpretive, and responsive essays for personal expression.

Proficient Level: Listening
- Students interpret important features of oral English language, at their grade level, relating to social academic topics and can discriminate between what is and what is not relevant.
- Students distinguish, conceptually or linguistically, complex oral English language expected of their grade level of fluent and/or native English speakers.
- Students comprehend grade-level English vocabulary, idioms, colloquial expressions, and apply their prior knowledge to grasp complex ideas expressed in English.
- Students listen to spoken English for a variety of purposes, including to acquire information and to take notes.

Proficient Level: Speaking
- Students select precise and descriptive grade-level vocabulary to participate actively in both social and academic English language settings.
- Students make use of standard English to communicate their ideas effectively in an organized and cohesive manner by adjusting to the social context to make themselves understood in English.
- Students utilize a variety of oral standard English language resources to analyze, solve problems, make decisions, and communicate shades of meaning in English.
- Students use oral standard English language to acquire, interpret, apply, and transmit information.
- Students present, in oral standard English language, their opinions and judgments on experiences, ideas, information, and issues.
- Students use the English language for effective social communication in socially and culturally appropriate manners.

Because there are two composite scores, the state has adopted a rule for determining proficiency from the two composites. That is, the overall proficiency level is defined by the lower of the two proficiency level designations. For example, if a student scores in the advanced level for listening/speaking and the proficient level for reading/writing, the overall level is advanced (CTB/McGraw-Hill, 2006).

Standard setting was based on the item mapping procedure (Mitzel, et al., 2001). The technical manual for 2006 indicates that detailed descriptions for each performance level exist, but they are not included in the manual.

Reliability and Validity

Information about the technical qualities of the assessment is provided in the technical reports.[13] Technical reports are prepared each year by the contractor, Pearson. The technical reports contain detailed information about test specifications, item and form development, item and form analysis, equating, and standard setting. They also contain results of analyses to evaluate reliability and validity, and they document efforts to evaluate fairness issues (e.g., bias review panels, analyses of differential item functioning). Reliability analyses include the standard types of analyses used for tests with multiple-choice items (i.e., estimates of internal consistency) as well as those used for open-ended items (i.e., interrater agreement), along with analyses of classification accuracy. Some validity evidence has been collected. Construct-related validity evidence was obtained by examining the intercorrelations between subtest scores and by conducting confirmatory factor analyses of the internal structure of the test. Evidence of criterion-related validity was collected by examining the degree of correspondence between students' performance on the NYSELAT and performance on the state's English assessments: for the lower grades, the latter was the state's English language arts assessment used for NCLB accountability purposes; for the higher grades, it was the Regents English exam.

STANFORD ENGLISH LANGUAGE PROFICIENCY TEST

The Stanford English Language Proficiency Test (SELP) was developed by NCS Pearson (formerly Harcourt Assessment, Inc.). Pearson offers both an "off-the-shelf" version of the assessment and customized versions that are augmented to meet a particular state's needs. Often the customized versions have different names. For instance, Arizona's version of the SELP is called the Arizona English Language Learner Assessment (AZELLA), and Washington's version is called the Washington Language Proficiency Test-II (WLPT-II). Currently, these are the only two user states, although when our study began, New Mexico and Wyoming also used the assessment (Roger Frantz, manager with Pearson, personal communication, June 2009).[14]

Content Standards

The test framework was developed in 1997 through analyses of the standards in place for six states (California, Delaware, Georgia, Hawaii, Missouri, and Texas), in conjunction with the TESOL standards (Roger Frantz, manager with Pearson, personal communication, June 2009). Frantz indicated that when a state chooses to

---

[13]The reports are available at http://www.emsc.nysed.gov/osa/reports/ [December 2010].

[14]Basic information about the assessment is available through Pearson at http://www.pearsonassessments.com/haiweb/cultures/en-us/productdetail.htm?pid=015-8429-206 [December 2010]. Information is also available at the websites for user states: for Washington, at http://www.k12.wa.us/assessment/wlptii/default.aspx [December 2010]; for Arizona, at http://www.ade.state.az.us/oelas/AZELLA/AZELLAAZ-1TechnicalManual.pdf [December 2010].

use the SELP, an alignment study is conducted to determine the extent to which the assessment is aligned with the state's ELP content standards. The test is then customized or augmented to ensure that the items cover the state standards.

Grade Bands

The SELP provides versions for six grade bands: Pre-K, K-1, 1-2, 3-5, 6-8, and 9-12. However, the grade bands can be customized for a state. For instance, Washington uses versions for four grade bands, K-2, 3-5, 6-8, and 9-12, and Arizona uses versions for five grade bands, K, 1-2, 3-5, 6-8, and 9-12.

Item Types

The SELP consists of five subtests: (1) listening, (2) reading, (3) writing, (4) writing conventions, and (5) speaking. The listening and reading subtests use multiple-choice items. The speaking subtest uses constructed-response items (described by the developer as "performance-based"). The writing conventions subtest uses multiple-choice items to measure the mechanics of writing. The writing subtest uses extended-answer constructed-response items.

Scores Reported

The off-the-shelf version of SELP offers scores for listening, speaking, and reading. Writing is a composite of the writing and writing conventions subtest. Five other composite scores are available: (1) productive skills [speaking and writing]; (2) comprehension skills [listening and reading]; (3) oral skills [listening and speaking]; (4) academic skills [reading, writing, and writing conventions]; and (5) an overall composite score. Washington reports individual domain scores (listening, speaking, reading, writing) and an overall composite score. Arizona reports the four domain scores and three composites (comprehension, oral, and overall composite).

Performance Levels

The off-the-shelf version of the SELP has set five performance level descriptions, but states are free to determine their own levels (Roger Frantz, manager with Pearson, personal communication, June 2009). The off-the-shelf version uses the following performance levels: pre-emergent, emergent, basic, intermediate, and proficient. The recommended cut scores for these levels were set by the publisher using the modified Angoff procedure (Angoff, 1984, also see Stephenson, 2003). For states that use a customized version, separate standard setting is done, and performance levels are adapted to the state needs.

Arizona uses the performance level names established for the off-the-shelf version (Porta and Vega, 2007, p. 137). The cutoff scores for the performance levels were determined through a standard setting based on the modified-Angoff procedure (Angoff, 1984; Reckase, 2000, as cited in Harcourt, 2007). Performance-level descriptions were developed for each domain area and for each grade band: that is, there are 20 sets of descriptions for the five performance levels. No overall performance-level descriptions appear to be available. As a sample of the performance level descriptions used by Arizona, below are the descriptions for the composite score in comprehension (reading and listening) for the middle elementary grades (3-5) (Harcourt, 2007).

Pre-Emergent: This student made very few or no responses. This student has very little ability to understand spoken English and understands only a few isolated words. This student understands almost no written English or only a few isolated words. This student may be able to understand visual universal symbols and graphics associated with a text.

Emergent: This student is able to comprehend a few key words, phrases, and short sentences in simple conversations on topics of immediate personal relevance when spoken slowly with frequent repetitions and contextual clues. This student is able to understand a few common high-frequency sight words and simple sentences in English. This student is able to comprehend a few simple content-area words with the aid of picture cues. This student is able to indicate the meaning of some common signs, graphics, and symbols.

Basic: This student is able to comprehend and follow three- to four-step oral directions related to the position of one's movements in space. This student can comprehend a few content-area words, including grade-level math and science vocabulary. This student is able to understand a few words that indicate mathematics operations. This student is able to comprehend some simple grade-level math word problems. This student comprehends and follows up to five-step written directions for classroom activities.

Intermediate: This student is able to comprehend and follow three- to four-step oral directions related to the position, frequency, and duration of one's movements in space. This student can comprehend some content-area words, including grade-level math and science vocabulary. This student is able to understand some words that indicate mathematics operations. Occasionally, this student is able to comprehend grade-level math word problems. This student comprehends and follows a short set of written instructions on routine procedures.

Proficient: This student comprehends and follows multiple-step oral instructions (four or more steps) for familiar processes or procedures. This student can comprehend many content-area words, including grade-level math and science vocabulary. This student is able to understand many words that indicate mathematics operations. Sometimes this student comprehends grade-level math word problems. This student comprehends and follows a set of written multi-step instructions on routine procedures.

Washington uses four performance levels: beginning/advanced beginning, intermediate, advanced, and transitional. Students must reach the transitional level to be considered for reclassification (Kimberly Hayes, WLPT-II memo, Office of Superintendent of Public Instruction, available: http://www.k12.wa.us/assessment/wlptii/pubdocs/WLPTMemoUpdated2010.pdf). Performance level descriptions are not provided in the technical manual but were obtained through the state Title III director (Helen Malagon, personal correspondence, September 2010):

Beginning/Advanced Beginning: Has little or no English reading skills with some understanding of content-area vocabulary and concepts. Writes simple English words, patterned phrases, and simple sentences. Communicates with words, sentences, drawings, gestures, and actions.

Intermediate: Comprehends short connected texts with context clues. Writes simple sentences or repetitive language. Participates in social discussions on unfamiliar topics. Begins to self-correct speech.

Advanced: Reads both short and long connected texts with understanding. Writes simple essays with standard conventions, organization, and detail. Uses figurative and idiomatic language in discussions of academic content and ideas.

Transitional: Reads and writes at grade level. Uses grammatically correct English with native-like proficiency.

Details about Washington's standard-setting methods are not described in the technical manual.

Reliability and Validity

Information about the technical qualities of the SELP assessment is provided in technical reports, some of which are available through state websites and some through the publisher.[15] The technical report for the WLPT-II was obtained from the state Title III director (Pearson Education, 2010). Technical reports do not ap-

---

[15]For instance, we obtained a technical report for the off-the-shelf version of SELP through the publisher, and we obtained the technical report for AZELLA at http://www.ade.state.az.us/oelas/AZELLA/AZELLAAZ-1TechnicalManual.pdf [December 2010]. We also obtained a technical report for New Mexico, for the 2007-2008 school year, at http://www.ped.state.nm.us/AssessmentAccountability/procurementLib3.html [December 2010]. We do not provide details about the New Mexico test because the state discontinued its contract with Pearson in 2009 and began using the ACCESS.

pear to be prepared each year for each state. An updated version of the technical report for the off-the-shelf version was still under preparation for the 2009 administration year. The version of Arizona's technical manual that the panel obtained was a summary of technical information for the 2006 administration year. The version of Washington's technical manual that we reviewed was for the 2008-2009 testing year.

The technical reports contain detailed information about test specifications, item and form development, item and form analysis, equating, and standard setting. They also contain results of analyses to evaluate reliability and validity and document efforts to evaluate fairness issues. For the SELP, reliability analyses include the standard types of analyses used for tests with multiple-choice items (i.e., estimates of internal consistency) as well as those used for open-ended items (i.e., interrater agreement). Studies of classification accuracy are also reported in the technical manuals for the two user states (Arizona and Washington). A number of validity studies have been conducted for both states. Evidence of content-related validity is based on studies of the alignment between the test items and the content standards. Evidence of construct-related validity is based on examination of the intercorrelations among the subtests, point biserial correlations, and principal components factor analyses of the internal structure. No evidence of criterion-related validity is reported for either state, although the report for Arizona indicates that such studies were planned for the 2007 testing cycle. Studies of fairness/bias appear to be based on bias reviews conducted as items were developed and test forms assembled. Results from analyses of differential item functioning are reported in the technical manual for Washington but not for Arizona.

TEXAS ENGLISH LANGUAGE PROFICIENCY ASSESSMENT SYSTEM

In response to state legislation passed in 1995, the Texas Education Agency (TEA), along with the testing contractor, Beck Evaluation and Testing Associates, developed the Reading Proficiency Tests in English (RPTE), which were implemented during the 1999-2000 school year for ELL students in grades 3 through 12. These were the first state-administered reading tests of ELP in the Texas assessment program. In response to federal requirements for assessing additional grades and language domains, additional assessments of English language proficiency were implemented during the 2003-2004 school year. At that time, the Texas English Language Proficiency Assessment System (TELPAS) was created, and RPTE was retained as the reading component of TELPAS for ELL students in grades 3-12. Holistically rated assessments were developed for the domain of reading in K-2 and for listening, speaking, and writing in K-12. Changes were made to the RPTE during the 2007-2008 school year, and the name RPTE was discontinued. The current version of the test is an online assessment. Technical information is available in the technical digest published by the Texas Education Agency (2009c).[16]

[16]Technical Digest 2008-2009, Chapter 7, TELPAS pp. 165-167; it is available at http://www.tea.state.tx.us/index3.aspx?id=2147484418&menu) [December 2010].

Content Standards

The RPTE was originally intended to align with the state's previous assessment program, the Texas Assessment of Academic Skills. Beginning in spring 2004, the RPTE was augmented in order to align it with another assessment, the Texas Assessment of Knowledge and Skills reading selections and test questions. In 2008, a new edition of RPTE was developed to align with the state's revised ELP standards, at which point a number of test modifications were made.

- The TELPAS subcomponent name RPTE was discontinued.
- A grade 2 test was added, resulting in the discontinuation of the previously administered holistically rated grade 2 TELPAS reading assessment.
- The grade clustering of the middle and high school tests changed from grades 6-8 and 9-12 to grades 6-7, 8-9, and 10-12.
- More reading selections and test questions were added to assess English language reading proficiency in mathematics and science contexts.
- The test blueprints were modified to include more reading material at the highest ELP level.
- The tests were developed as online assessments.

TELPAS is intended to measure learning in alignment with the Texas English Language Proficiency Standards, which are a component of the Texas Essential Knowledge and Skills curriculum. The standards outline the instruction that ELL students must receive to support their ability to develop academic ELP and acquire challenging academic knowledge and skills.

Grade Bands

The TELPAS reading tests have versions for the following grade bands: 2, 3, 4-5, 6-7, 8-9, and 10-12. The holistically rated components are grade specific.

Item Types

TELPAS includes holistically rated, performance-based components to assess skills in some of the domains. These assessments are used in all domains for grades K-1, listening, speaking, reading, and writing. For grades 2-12, they are used to assess all domains except reading, which is assessed through multiple-choice items. The holistic assessments are conducted by teachers in the classroom. The teachers are trained to collect information on their own students and to evaluate on the basis of their interactions with and observations of students. Writing in grades 2-12 is assessed through a collection of students' classroom writing assignments. Teachers must undergo training to learn how to conduct the ratings and must meet qualification standards. The rating rubrics are the proficiency-level descriptors, which are defined in the Texas ELP standards and which teachers are required to use in ongoing instruction to develop students' English language proficiency and make grade-level instruction linguistically accessible.

Scores Reported

Scores are reported for each domain, listening, speaking, reading, and writing. Two composite scores are also reported. One is a comprehension score, derived from performance on the reading and listening subtests. An overall composite score and rating are also reported. In computing this composite score, listening and speaking are each weighted by 5 percent, writing is weighted by 15 percent, and reading is weighted by 75 percent. According to the technical manual (Texas Education Agency, 2009c) listening and speaking receive less weight so that students do not attain a high composite proficiency rating before they acquire the English reading and writing proficiency needed to support their full potential for academic success.

Performance Levels

TELPAS scores are reported according to four performance levels: beginning, intermediate, advanced, and advanced high. Performance-level descriptions are available for each domain and for the overall score. The global descriptors appear below:

> *Beginning:* Beginning students have little or no ability to understand and use English. They may know a little English but not enough to function meaningfully in social or academic settings.
>
> *Intermediate:* Intermediate students do have some ability to understand and use English. They can function in social and academic settings as long as the tasks require them to understand and used simple language structures and high-frequency vocabulary in routine contexts.
>
> *Advanced:* Advanced students are able to engage in age-appropriate academic instruction in English, although ongoing second language support is needed to help them understand and use grade-appropriate language. These students function beyond the level of simple, routinely used English.
>
> *Advanced High:* Advanced high students have attained the command of English that enables them, with minimal second language acquisition support, to engage in regular, all-English, academic instruction at their grade level.

To be considered proficient in English on the TELPAS, students must score at the "advanced high" level.

Reliability and Validity

Information about the technical qualities of the assessment is provided in annually published technical digests.[17] The technical digests are prepared for each administration cycle by TEA, in conjunction with Pearson, the state's testing contractor. The version used for the panel's review was for the 2008-2009 school year (Texas Education Agency, 2009c). This digest contains detailed information about test specifications, item and form development, item and form analysis, and statistical procedures for equating of the reading test. The holistically rated assessments are not statistically equated; instead, the difficulty is maintained through the use of consistent rating rubrics developed to define the proficiency levels and through consistent training and qualifying procedures for the raters. Details about standard setting appear in the report for the 2007-2008 school year.

The technical report contains results of analyses to evaluate reliability and validity. Reliability analyses include the standard types of analyses used for tests with multiple-choice items (i.e., estimates of internal consistency) as well as those used for open-ended items (i.e., interrater agreement). Estimates of classification accuracy are also provided (e.g., accuracy of student classifications into performance categories). Some validity evidence has been collected. Content-related validity evidence consists primarily of expert review of the extent to which the items correspond/conform to the item specifications and the performance-level descriptions. The TEA indicates that construct-related validity evidence is provided through estimation of internal consistency reliability for the multiple-choice components and the training and administration procedures for the holistically rated components. Evidence of criterion-related validity was collected by examining the degree of correspondence between performance on the TELPAS reading component and performance on the state's reading assessment (the Texas Assessment of Knowledge and Skills, TAKS). For the study, the average TAKS reading score was calculated for students at each grade level and at each performance level: for example, the mean TAKS score for 3rd graders classified on the TELPAS as beginning, intermediate, advanced, or advanced high, and so on for each grade). Rating audits of the other language domains are conducted to provide evidence that the internal structure of the assessments are intact and that teachers administer the holistically rated assessments and apply the rating rubrics as intended. No information is provided about attempts to evaluate the assessment for fairness or bias.

[17]The digests are available at http://www.tea.state.tx.us/index3.aspx?id=2147484418&menu_id=793) [December 2010].

Appendix B

Biographical Sketches of Panel Members and Staff

Alan M. Zaslavsky (*Chair*) is professor of statistics in the Department of Health Care Policy, Harvard Medical School. His statistical research interests include surveys, census methodology, small-area estimation, official statistics, missing data, hierarchical modeling, and Bayesian methodology. His research topics in health care policy focus on measurement of the quality of care provided by health plans through consumer assessments and clinical and administrative data. Currently, his major projects include survey implementation for the Medicare system, methodology for surveys in psychiatric epidemiology, and studies on determinants of quality of care for cancer. Other research interests include measurement of disparities in health care, and privacy and confidentiality for health care data. He is a fellow of the American Statistical Association. He received an A.B. from Harvard College, an M.S. from Northeastern University, and a Ph.D. from the Massachusetts Institute of Technology.

Jamal Abedi is professor of education in the School of Education at the University of California at Davis. Previously, he was a faculty member at the Graduate School of Education of the University of California at Los Angeles and director of technical projects at the Center for Research on Evaluation, Standards, and Student Testing. His research interests include psychometrics and test and scale development. He also conducts research on the use of latent-variable modeling to assess the validity and reliability of performance-based assessment measures. His recent work includes validity studies for the National Assessment of Educational Progress (NAEP), focusing on the impact of language background on students' performance and the dimensionality of NAEP mathematics subscales. He has also developed a culture-free instrument for measuring creativity, which has become translated into a number of languages and administered in several countries. He holds a Ph.D. in psychology from Vanderbilt University.

Frank Bean is chancellor's professor in the School of Social Sciences and director of the Center for Research on Immigration, Population, and Public Policy at the University of California at Irvine. Previously, he served as Ashbel Smith professor of sociology and public affairs and director of the Population Research Center at the University of Texas at Austin. He has been a visiting scholar at the Australian National University, the American Academy in Berlin, and the Russell Sage Foundation, as well as distinguished senior visiting fellow at the College Consortium for International Studies and the Center for U.S.-Mexico Relations at the University of California at San Diego. His current research focuses on the implications of U.S. immigration policies, Mexican immigrant incorporation, the implications of immigration for changing race/ethnicity in the United States, the determinants and health consequences of immigrant naturalization, and the development of new estimates of unauthorized immigration and emigration. He is a member of Phi Beta Kappa, Phi Kappa Phi, and the Council on Foreign Relations. He holds a Ph.D. from Duke University.

David Francis is Hugh Roy and Lillie Cranz Cullen distinguished professor and chair of the Department of Psychology at the University of Houston and a recipient of the university's Teaching Excellence Award. His areas of quantitative interest include modeling of individual growth, multilevel and mixture modeling, structural equation modeling, item response theory, and exploratory data analysis. His current research includes work supported by the National Institute of Child Health and Human Development, the Institute of Education Sciences, the National Institute of Deafness and Communication Disorders, the Texas Education Agency, and the Houston Livestock Show and Rodeo. He is a fellow of Division 5 (measurement, evaluation, and statistics) of the American Psychology Association. He holds a Ph.D. in clinical neuropsychology from the University of Houston.

Edward Haertel is Jacks Family professor and associate dean for faculty affairs at the School of Education at Stanford University. His research centers on policy uses of achievement test data; the measurement of school learning; statistical issues in testing and accountability systems; and the impact of testing on curriculum and instruction. He has been closely involved in the creation and maintenance of California's school accountability system both before and after passage of the No Child Left Behind Act and has served on advisory committees for other states and for testing companies. In addition to technical issues in designing accountability systems and quantifying their precision, he is concerned with validity arguments for high-stakes testing, the logic and implementation of standard setting methods, and comparisons of trends on different tests and in different reporting metrics. He has served on numerous state and national advisory committees related to educational testing, assessment, and evaluation, including the committee responsible for the 1999 revision of the *Standards for Educational and Psychological Testing*. He is a fellow of the American Psychological Association and is a member of the National Academy of Education. He holds a Ph.D. from the University of Chicago.

David Hubble is a senior statistician at WESTAT with extensive experience in designing, planning, and conducting demographic surveys and census evaluations at the U.S. Census Bureau. At WESTAT, his work has involved the National Assessment of Educational Progress, the Minnesota Adult Tobacco Survey, and other survey design and technical assistance projects. His work covers many aspects of survey implementation, including survey design, sampling frame creation, sample selection, data collection methods, missing data mitigation, weighting procedures, estimation techniques, variance estimation, methodological investigations, and experimental designs. He holds a B.A. and an M.A. in statistics, both from Boston University.

Judith A. Koenig (*Costudy Director*) is a senior program officer with the Board on Testing and Assessment of the National Research Council (NRC). She has worked on a number of projects related to assessing English language learners, as well as an evaluation of the assessments used by the National Board for Professional Teaching Standards, a project to recommend standards for the National Assessment of Adult Literacy, and a report on methods for calculating dropout and graduation rates. Prior to joining the NRC, she was a senior research associate with the Association of American Medical Colleges where she directed operational programs for the Medical College Admission Test and led a comprehensive research program on the examination. She has a B.A. in special education from Michigan State University, an M.A. in psychology from George Mason University, and a Ph.D. in educational measurement, statistics, and evaluation from the University of Maryland.

Rebecca Kopriva is a visiting professor at University of Wisconsin-Madison, working with the Wisconsin Center on Educational Research. Previously, she was on the faculty at the University of Maryland, where she served as director for the Center for the Study of Assessment Validity and Evaluation (CSAVE) in the College of Education. At CSAVE, her work focused on improving the quality and accuracy of data about students and schools, with a particular focus on addressing ways large scale psychometric theory and practice can be improved to better incorporate the needs and strengths of diverse student populations. Prior to that position, she served as director of student assessment for the state of Delaware. Much of her research focuses on the issue of assessing English language learners, and she recently served as consultant on the English Language Development Assessment on the validity of the assessments. She holds a Ph.D. in applied statistics and research methods from the University of Northern Colorado.

Robert Linquanti is project director for English Learner Evaluation and Accountability Support and senior researcher for the Regional Educational Laboratory West and the California Comprehensive Center at WestEd. He specializes in assessment, evaluation, and accountability policies and practices and systems for English language learners. He regularly serves as a consultant on the assessment of English language learners and accountability policy and practice issues to the Council of Chief State School Officers; the 23-state World-Class Instructional Design and Assessment Consortium; the U.S. Department of Education; and the National Assessment Governing Board. He leads WestEd's multiyear collaboration with the California Department of Education and regional service providers to deliver technical assistance for school districts identified under Title III as needing to improve educational services and outcomes for English language learners. He holds a B.A. in English and Spanish literature and linguistics from the State University of New York at Buffalo and an M.P.A. in public policy analysis from Columbia University.

Helen Malagon is the interim director of Bilingual and Migrant Programs for the Office of Superintendent of Public Instruction in Washington State. In this capacity, she directs and manages all aspects of the state's Transitional Bilingual Instruction Program, Title III, and Title I Part C-Migrant Education. Her primary responsibility is to interpret state and federal legislation as it pertains to English language learners and migrant programs and to monitor school districts for compliance with state and federal requirements. She also leads the state's Bilingual Education Advisory Committee, whose role is to provide guidance to the state regarding the needs of English language learners. Previously, she was the coordinator for curriculum and instruction with the High School Equivalency Program for migrant students at the University of New Mexico and director of the Title VII programs for Native Americans in Nebraska and in South Dakota.

Catherine Neff is coordinator of the Title III/English for Speakers of Other Languages (ESOL) program for the South Carolina Department of Education. In this position, she interprets federal statutory and regulatory requirements relative to the administration and implementation of Title III at both the state and local levels and administers Title III grants and monitors districts for compliance with Title III law. She also provides professional assistance to districts in serving the needs of limited English proficient and immigrant students and their families and provides statewide assistance to ESOL teachers, mainstream teachers of ESOL students, and administrators in the development and implementation of services for English language learners. She serves on the Hispanic Advisory Board for the South Carolina Commission on Minority Affairs and is a member of Clemson University's Hispanic Task Force. Previously, she administered a credit recovery program for high school at-risk students and English language learners in Salinas, California.

P. David Pearson is dean of the Graduate School of Education and professor of language and literacy, society, and culture at the University of California at Berkeley. His research interests include practice and policy in literacy instruction and assessment. Previously, he was dean of the College of Education of the University of Illinois at Urbana-Champaign, where he also codirected the Center for the Study of Reading, and the John A. Hannah distinguished professor of education at Michigan State University. He has served as president of the National Reading Conference and on the boards of directors for the International Reading Association, the National Reading Conference, and the Association of American Colleges of Teacher Education. He is a member of the National Academy of Education and a recipient of the William S. Gray Citation of Merit from the International Reading Association, the Oscar Causey Award for Contributions to Reading Research from the National Reading Conference, and the Alan Purves Award from the National Council of Teachers of English. He holds a B.A. in history from the University of California at Berkeley and a Ph.D. in education from the University of Minnesota.

Thomas Plewes (*Costudy Director*) is a senior program officer for the Committee on National Statistics of the National Research Council, and he served as study director for the Panel to Review Research and Development Statistics at the National Science Foundation. Previously, he was associate commissioner for employment and unemployment statistics of the Bureau of Labor Statistics, and he served as chief of the U.S. Army Reserve. He is a fellow of the American Statistical Association and was a member of the Federal Committee on Statistical Methodology. He has a B.A. in economics from Hope College and an M.A. in economics from the George Washington University.

Charlene Rivera was the founder and is the executive director of the George Washington University's Center for Equity and Excellence in Education (CEEE). The CEEE's goal is to advance education reform in a way that supports equity while enhancing the achievement of all students. The portfolio of work carried out by the center includes policy research in areas such as the assessment of English language learners, providing technical assistance in education reform to state departments of education, districts, and schools and conducting program evaluations for school districts. She has published extensively on the issue of assessing English language learners and led a project that generated tools for policy makers, educators, and community members to help English learners reach high academic standards. She currently serves on the technical committee for the World-Class Instructional Design and Assessment Consortium that is developing English language proficiency tests for 10 partner states. She is formerly a teacher for the Boston Public Schools. She holds an Ed.D. from Boston University.

COMMITTEE ON NATIONAL STATISTICS

The Committee on National Statistics (CNSTAT) was established in 1972 at the National Academies to improve the statistical methods and information on which public policy decisions are based. The committee carries out studies, workshops, and other activities to foster better measures and fuller understanding of the economy, the environment, public health, crime, education, immigration, poverty, welfare, and other public policy issues. It also evaluates ongoing statistical programs and tracks the statistical policy and coordinating activities of the federal government, serving a unique role at the intersection of statistics and public policy. The committee's work is supported by a consortium of federal agencies through a National Science Foundation grant.

BOARD ON TESTING AND ASSESSMENT

The Board on Testing and Assessment (BOTA) assists policy makers and the public by providing scientific expertise around critical issues of testing and assessment in education, the workplace, and the armed services. BOTA's fundamental role is to raise questions about—and provide guidance for judging—the technical qualities of tests and assessments and the intended and unintended consequences of their use. BOTA consists of experts from a range of disciplines relevant to testing and assessment—psychology, statistics, education, economics, law, business, anthropology, sociology, and politics—as well as practitioners with experience in test use. Among the issues BOTA considers are the uses of tests as policy tools, civil rights implications of tests, and innovative methods of assessment.